Values and the
Environment

Values and the Environment
A SOCIAL SCIENCE PERSPECTIVE

Edited by
YVONNE GUERRIER
Department of Hospitality, Food and Product Management, South Bank University, UK

NICHOLAS ALEXANDER
School of Commerce and International Business Studies, University of Ulster, UK

JONATHAN CHASE
Department of Psychology, University of Surrey, UK

MARTIN O'BRIEN
Department of Sociology, University of Surrey, UK

JOHN WILEY & SONS
Chichester · New York · Brisbane · Toronto · Singapore

Other Wiley Editorial Offices

John Wiley & Sons, Inc., 605 Third Avenue,
New York, NY 10158-0012, USA

Jacaranda Wiley Ltd, 33 Park Road, Milton,
Queensland 4064, Australia

John Wiley & Sons (Canada) Ltd, 22 Worcester Road,
Rexdale, Ontario M9W 1L1, Canada

John Wiley & Sons (SEA) Pte Ltd, 37 Jalan Pemimpin #05-04,
Block B, Union Industrial Building, Singapore 2057

Library of Congress Cataloging-in-Publication Data
Values and the environment : a social science perspective / edited by
    Yvonne Guerrier ... [et al.].
        p.   cm.
    Includes bibliographical references and index.
    ISBN 0-471-96047-0 (alk. paper)
    1. Environmental ethics. 2. Environmental policy. 3. Social
values. I.   Guerrier, Yvonne.
GE42.V35   1995
179'. 1–dc20   95-16830
                                                              CIP

British Library Cataloguing in Publication Data

A catalogue record for this book is available from the British Library

ISBN 0 471 96047 0

Typeset in 10/12pt Times by Saxon Graphics Ltd, Derby
Printed and bound in Great Britain by Bookcraft (Bath) Ltd, Midsomer Norton
This book is printed on acid-free paper responsibly manufactured from sustainable forestation,
for which at least two trees are planted for each one used for paper production.

Contents

Contributors

Nicholas Alexander is Senior Lecturer in Retailing in the School of Commerce and International Business Studies, Faculty of Business and Management, University of Ulster. He was previously Lecturer in Retail Management at the University of Surrey and Coca Cola Lecturer in Retailing at the University of Edinburgh. His primary research interest is the internationalisation of retailing. His work has been presented at conferences, and has appeared in journals, in Europe and North America. He is an associate editor of *The Service Industries Journal*.

Marino Bonaiuto is a Research Fellow in the Dipartimento di Psicologia dei Processi di Sviluppo e Socializzazione, Università degli Studi di Roma 'La Sapienza' and in the Department of Psychology at the University of Surrey, Guildford. He has been involved in research on the perception of urban environments and on the perception of pollution on beaches.

Mirilia Bonnes is Professor in the Dipartimento di Psicologia dei Processi di Sviluppo e Socializzazione, Università degli Studi di Roma 'La Sapienza'. She has undertaken research on individuals' perceptions of urban environments for UNESCO. She has also researched the organisation of personal space in different cultures and is presently working on a book of social psychological perspectives on environmental psychology.

Kate Burningham is Lecturer in Sociology of the Environment in the Department of Sociology and the Centre for Environmental Strategy at the University of Surrey. Her current research is on the social construction of local environmental impacts.

Jonathan Chase is a Lecturer in the Department of Psychology at the University of Surrey. He is involved in EU-funded research on the social and psychological determinants of environmental behaviour.

Judy Clark is a Lecturer in the Department of Geography, University College, London.

Roland Clift is Professor of Environmental Technology and Director of the Centre for Environmental Strategy at University of Surrey. He is a chemical engineer by profession. His current work centres on the development and use of environmental life cycle assessment techniques.

Dominic Dibble is a Teaching Assistant in Educational Policy and Development at the University of Stirling. He researches the use of computers and information technology in the development of education.

Michael John Doupé is Lecturer in Law at the University of Lancaster. He has lectured widely to both the legal profession and businesses on the regulation of environmental pollution and is co-author of a forthcoming book on business environmental regulation.

Yvonne Guerrier co-ordinated the organisation of a conference on 'Values and the Environment' when she was a Senior Lecturer in the Department of Management Studies in the University of Surrey. She is now Professor of Hotel Management at South Bank University, London.

Martin J. Haigh is Professor of Physical Geography at Oxford Brookes University where he teaches courses on soil conservation, environmental geomorphology, gaia, and environmental philosophy. His major research interests concern the practical reconstruction of damaged landscapes – especially in coal mining and mountain regions in Wales, the Himalayas and Eastern Europe – and educational methods. Haigh is current Vice President of the World Association of Soil and Water Conservation. Before 'de-emigrating' to Oxford, he worked at the University of Chicago and with the American Institute of Indian Studies.

Ragnar Löfstedt is a Lecturer in Social Geography working on energy and environmental problems mainly from a risk perception/communication perspective in Sweden, the Baltic States and the UK. He holds a BA in geography from UCLA and an MA and PhD in geography from Clark University in the United States.

Merylyn McKenzie Hedger is Research Fellow at the Institute of Earth Studies, University of Wales, Aberystwyth. She has worked in public administration and as a consultant on a wide range of environmental planning issues in the UK and overseas.

Martin O'Brien is a Lecturer in Sociology at the University of Surrey. He works in the areas of social theory, social change and sociology of the environment, among others. He is currently developing research in the area of integrated methods for social and environmental life cycle assessment with Roland Clift from the Centre for Environmental Strategy, University of Surrey.

Ioannis S. Panagopoulos is a Researcher in the Department of Psychology at the University of Surrey. He is involved in EU-funded research on the social and psychological determinants of environmental behaviour.

Jenneth Parker is a part-time Tutor at the Centre for Continuing Education at the University of Sussex and a Consultant in Community Education for

Sustainability. She is studying for a DPhil in Ecofeminist Ethics at the Women's Studies Graduate Division, University of Sussex.

Peter J. G. Pearson is Principal Research Fellow at the Centre for Environmental Technology, Imperial College of Science, Technology and Medicine, London. He was previously Director of Surrey Energy Economics Centre at the University of Surrey. In 1993 he held a UK Economic and Social Research Council Global Environmental Change Research Fellowship.

Michael R. Redclift was Director of the Global Environmental Change Programme of the Economic and Social Research Council between 1990 and 1995. He is a member of the Environmental Section at Wye College, where he specialises in environment and development issues.

Adam Rutland is a British Academy Post-doctoral Research Fellow in the Department of Psychology, University of Surrey. His main area of research is in developmental and social psychology. Recently he worked with David Uzzell as a Research Fellow on an EU-funded environmental education project.

Mick Smith is a 'lay academic' with a background in ecology, philosophy and social theory. He currently teaches environmental sociology in the Department of Applied Social Science at the University of Stirling. He lives with his partner and a cat in a farm cottage surrounded by hares, swallows and skylarks.

David L. Uzzell is Senior Lecturer in Psychology and Course Director for the MSc in Environmental Psychology, University of Surrey. He is a social and environmental psychologist whose research includes urban planning, safety and the environment, heritage and environmental interpretation, and museum visitor studies. He recently co-ordinated a four-nation research project for the Commission of the European Communities on 'children as catalysts of global environmental change'.

David Whistance spent eight months working in a resource development project run by the Wildlife Society of Southern Africa after graduating with a psychology degree from the University of Surrey in 1993. He intends to train as a careers guidance officer.

Acknowledgements

Our special thanks to the Faculty of Human Studies at the University of Surrey for financial, moral and organisational support and to all those who helped organise the 'Values and the Environment' conference, notably Janet Adshead-Lansdale, David Hawdon, Peter Pearson, Chris Flood, Mike Hornsby-Smith, Owen Rees, Anne Riggs and especially Jackie Abbott and Joan Leacock without whose administrative skills the conference would not have taken place. Many thanks to Andrew Yip for sterling work. Our thanks also to the contributors to this volume for meeting our deadlines and conditions and to Iain Stevenson at John Wiley for his support and advice.

Values and the Environment: An Introduction

MARTIN O'BRIEN AND YVONNE GUERRIER

> Not to put too fine a point on it, we live, breathe, and excrete values. No aspect of human life is unrelated to values, valuations and validations. Value orientations and value relations saturate our experiences and life practices from the smallest established microstructures of feeling, thought and behaviour to the largest established macrostructures of organisations and institutions. (Fekete 1988)

All individual and collective action is informed by values. These may be personal values which each actor holds: our motives, reasons or justifications for action. Conversely, they may be the values embedded in a social context (family, community, school or work, for example) which constrain or enable individual action whether they accord with personal values or not: values embodying ideals of respect, effort or obedience, for example. At the same time all such values are practically and concretely *realised in* social action and organisation. The effects of values on the world of experience only become tangible, interpretable or contestable when one action framework encounters another: when more than one way of acting in and on the world is possible and a choice must be made. The problematic relationship between action and values persists in the complex intersection of motives, constraints, contexts and choices that underpins both the ordinary conduct of everyday life and the systems of production, organisation and control that sustain wider societies.

This means that progress towards goals of environmental repair, sustainability or development cannot be achieved simply by encouraging the adoption of one set of abstract values over another, or by trying to convince people that one value is 'better' than another. Values circulate and mutate, are foregrounded or backgrounded, adopted or excluded on the basis of a very wide range of social, cultural, economic and political priorities and commitments. In many ways, it is the incommensurability of the priorities, the diversity of the commitments, the essential differentiation of value systems that renders problematic attempts to develop a normative ethics of environmental care and responsibility. What is 'ethical' and 'responsible' in one community is sacrilegious and capricious in another.

Values and the Environment: A Social Science Perspective, Edited by Yvonne Guerrier, Nicholas Alexander, Jonathan Chase and Martin O'Brien. © 1995 John Wiley & Sons Ltd.

Sometimes the opposition persists through the veneration of and commitment to distinctive collections of beliefs and attitudes. Today, archaeologists in America are in dispute with native Amerindians around ownership and rights of access to the ancient remains of the latters' ancestors. While the scientific view extols the 'value' of the large collection of bones for understanding human development and the mutation and spread of diseases, the Amerindian view upholds the spiritual 'value' of the remains and insists that they be returned for reburial in keeping with traditions far more ancient than the discipline of archaeology. In a case of paradigm conflict like this, the communal traditions which uphold one value as against another cannot be dismantled by contesting the rationality or utility of one belief or attitude from the perspective of an alternative one.

Sometimes the opposition is based on the real destructive consequences attendant on the imposition of alien value systems through political, cultural or economic colonisation or displacement; many of the intractable disputes about environmentally friendly practice (whether in disputes over road-building programmes, the dumping of toxins, the use of chemical fertilisers, the waste of energy resources, or the destruction of rainforests) can themselves be seen as conflicts over rights to define what is and is not valuable about environments and to have those definitions accepted and acted upon. Such conflicts often represent overt political struggles between different societal interest groups, where 'values' are mobilised as resources in order to achieve a variety of ends against a more (or less) powerful opposing group (see Burningham and O'Brien 1994).

Environmental values are today invariably understood as 'green' values: as values that propose or support action directed towards environmental care and responsibility. Yet, in reality, environmental values lock into economic values, political and intellectual values; they are not the sole province of 'green' thinkers but are dispersed among many different organisations reflecting very different moral and political standpoints. Regardless of the relative superiority of some value position or another, environmental change is occurring and will necessarily continue beyond the lifetimes of readers of this book. Similarly, the values attached to different environments will also change – into forms that today are literally unimaginable. The point, in short, is that there is no possibility that contemporary researchers, scholars or activists could develop a single value or value system able to encompass the myriad strains of belief, commitment and attitude that envelop people's relationships with their environments.

Values are important in the debate about the environment not because some value or other in itself can or should be described as 'right' or 'wrong', but because value systems refer to underlying principles about the 'proper conduct' of life in general and about ways of interpreting specific events in terms of more extensive commitments to particular social arrangements and political orders. They indicate the cultural plurality – and often ambiguity – within which notions of 'rightness' and 'wrongness' are formulated, maintained, contested and changed.

'Values and the environment', then, is a thorny nest of intellectual and political problems. It delineates a complex field where ideas and visions, rights and responsibilities encounter traditions and interests, institutions and technologies, all of which are essentially contested at the level of experience. Yet the complexity and contested nature of the field do not necessarily imply paralysis or powerlessness in the face of environmental change. Although there are many ways in which dominant approaches to environmental understanding or education are disempowering (see Chapters 4 and 12), there are equally as many opportunities for democratising or re-empowering people's relationships with the environment (see Chapters 3 and 14) – at the level of understanding and at the level of political action. Any meaningful engagement with the field must address the different reasons underlying and the different purposes served by processes of environmental valuation. Environmental valuations fulfil many uses: environments may be valued for the profit or economic benefit they can bequeath, they may be valued for the multiple pleasures that they bestow, they may be valued for the security or stability they provide in some people's lives. These and other types of valuation that are observed in this collection must be of equal concern to the policy community if goals of sustainability and environmental care are to gain sufficient support to make them realistic political and social targets.

The collection of essays in the present volume is motivated precisely by the ambiguity and ambivalence of the current environmental debate. It is not intended to provide quick fixes or clichéd solutions to pressing environmental concerns. Rather, it is intended to make available to a multidisciplinary audience a range of responses to key features of the problems that become visible when the environment is a focus of experiment, analysis, governance and action. These problems revolve around the historical emergence and development of dominant cultural symbols and discourses on the environment; around the intellectual paradigms in which environmental values are warranted and incorporated into social systems of knowledge production; around the political agenda pursued by institutional and communal interest groups; and around the practical effort to encourage the adoption of 'environmental values'. As readers will discover, these problems admit of many different solutions, none of which provide *the* answer to the dilemmas posed by the perception and experience of environmental change.

The collection derives from a conference on 'Values and the Environment' held at the University of Surrey in September 1993. All of the contributors to this volume were connected with the conference; some of the chapters were specially written for this book, while others are revised versions of conference papers. A particular aim of the conference was to provide a forum for debate where people from a wide range of academic disciplines, representing a diversity of interests and a variety of perspectives, could trade knowledge and expertise in assessing the relationships between individual and societal value systems and the environment. Further, the conference was intended to bring together

scholars and researchers who would not normally share the same public plat-
form or exchange their different perspectives on the processes and contexts of
environmental valuation; to bring together people whose primary research
focus was the environment and environmental change and people working on
other topics whose skills might contribute to clarifying the values–environment
connection; people with clear, personal commitments to certain sets of environmental values and those who were more agnostic in their value orientations. In
these senses, the conference was intended to generate debate and to enable an
open dialogue to take place between a wide range of academic and value con-
stituencies.

This book is inevitably more focused than the conference, yet it is informed
by these same 'values'. The contributors are drawn from backgrounds in
philosophy, sociology, psychology, economics, law, geography, education and
engineering in order to explore the many different ways that both environmental 'experts' and 'lay' analysts *engage in the process* of placing values on the
environment, rather than to promote a 'right' way of doing so. Some of the contributors have chosen to be definite in the expression of their values, others
reveal little explicitly of their own commitments and preferences; this contrast,
of course, also reflects the different theoretical approaches in the social sciences adopted by the various contributors.

The book is divided into four sections. The first section addresses fundamental theoretical and conceptual matters about the nature of values and the environment. Here, general issues of the character and ethical frameworks of
modern societies are examined together with the relationships between forms of
democracy, community empowerment and the maintenance of conflicting value
perspectives. The second section explores some of the conceptual and methodological problems and prospects involved in the process of evaluating people's
values. This complex doublet raises serious technical and political questions
both about the nature of values as such and the contexts and frameworks to
which they refer. The third section investigates some of the policy dilemmas
associated with environmental management and regulation. The chapters in this
section debate the relationships between national and international policy bodies, the role of law and the judicial system in environmental regulation, and the
tensions inherent in current regulatory frameworks as a result of the existence
in policies of multidimensional evaluative perspectives. The final section deals
with issues of value change through education about the environment. The contributors to this section challenge conventional views on the process of learning
about the environment and propose alternative ways forward in the field of educational provision for dealing with the persistence of competing values.

Each section begins with an independent introduction which establishes the
key themes underpinning the debates and the parameters within which they are
conducted. Although each section examines a different dimension of the values–environment nexus, several themes recur regularly throughout the analyses.
These include the tensions between local and global perspectives and actions,

the bewildering variety of interpretations provided by both expert and lay approaches to environments, the sources of environmental values and understandings, and the relationships between values and the contexts in which they are formed and changed.

This volume therefore represents an intervention into *the debate about environmental change* rather than a handbook of worked-out recommendations for action. In the chapters that follow, there are many practical suggestions, based on the authors' researches and experiences, about how the manifold assemblages of private and public value systems can be addressed positively in the development of environmental awareness and action. Yet, as a number of authors make abundantly clear, this development cannot occur solely at the level of values: the contexts of emergence and the conditions of persistence of those values are themselves crucially important targets for social and political action. If readers gain a sense of the inherent interconnectedness of values, actions and contexts, then one of the volume's aims will have been achieved. If, on the basis of this recognition, the environmental debate is better informed, more open to the plurality of value positions which can generate environmental awareness and wisdom, more positively critical of the relationships between beliefs, principles, actions and organisation, then it will be in a better position to promote democratic, rather than dogmatic, solutions to environmental change.

REFERENCES

Burningham, K. and O'Brien, M. (1994) 'Global Environmental Values and Local Contexts of Action', *Sociology*, **28**, 4, 913–32.

Fekete, J. (1988) 'Introductory Notes for a Postmodern Value Agenda' in J. Fekete (ed) *Life After Postmodernism: Essays on Value and Culture*, Macmillan, London, pi, cited in Steve Connor, 1993, 'The Necessity of Value', in J. Squires (ed) *Principled Positions: Postmodernism and the Rediscovery of Value*, Lawrence & Wishart, London.

Section I

DEBATES AND ISSUES

Introduction

YVONNE GUERRIER
South Bank University

This first section of the book aims to introduce the reader to some of the broad debates and issues about the valuation of the environment and the way values inform and should inform public action. The four chapters in this section explore a wide range of issues from the way in which the values of modernist science and politics inform the global environmental debate (Redclift) to the variety of approaches that may be taken when assessing the impact of wind turbines on the environment (Clift, Burningham and Löfstedt); from the extension of local democracy to include local autonomy in environmental decision making (Parker) to the need to break away from current moral paradigms when placing a value on the environment and to search for a genuinely oppositional moral philosophy (Smith). What the chapters share is a desire to confront fundamental questions about the valuation of the environment and to discuss these questions in the context of wider philosophical, sociological and scientific debates. What assumptions underlie thinking about the environment? What role can 'experts' play in the valuation of the environment and what paradigms do they work within? Can ordinary people be trusted with the care of the environment? What is the relationship between the valuation placed on the local environment and that placed on the global environment?

In the first chapter, Redclift focuses on values and *global* environmental change. He argues that global environmental change needs to be understood in relation to three sets of issues. First, he argues that the way science and social science have developed from the nineteenth century onwards has divorced the study of 'nature' from the study of society and made it particularly difficult to deal with those issues, such as environmental issues, which reflect the interface between society and nature. Secondly, he discusses the way in which the growth of scientific knowledge, instead of increasing our ability to predict and control the future, has drawn attention to the gaps in our knowledge and our inability to find technical fixes for global environmental problems. 'We know more but we are able to do less'. Finally he argues that, however much governments may espouse the values of sustainable development, practical politics and science have their base in a set of values that are unsustainable.

The second chapter is written by three authors (Clift, Burningham and Löfstedt) with backgrounds in engineering, sociology and geography respec-

Values and the Environment: A Social Science Perspective, Edited by Yvonne Guerrier, Nicholas Alexander,
Jonathan Chase and Martin O'Brien. © 1995 John Wiley & Sons Ltd.

tively. Taking wind power and the use of wind turbines as an example, it examines the way engineers traditionally conceptualise environmental assessment and provides a critique, from a social science perspective, of this approach. The paper discusses the problems of integrating the perspectives of various groups of 'experts' to inform the public debate on environmental issues, given that the concepts of 'system' and 'environment' may be understood in such different ways.

Parker's focus, in the third paper, is on the process of public debate about environmental issues. She contrasts an 'expert' model of incorporating environmental values into public policy, where the main need is to convince decision makers at a high level in the system to take into account ethical issues in relation to the environment, with a 'process' model where the aim is to develop morally reflective communities that can take stewardship over the local environment. She argues for an extension of local democracy to include greater autonomy in environmental decision making, claiming that this is ethically preferable to imposing values through central policy and that it ensures the 'ownership' necessary for values to be championed.

Smith's focus is also on environmental ethics. He challenges those 'experts', such as environmental economists and moral philosophers, who claim to have conceptual systems and methodologies to evaluate the 'natural' world, for failing to recognise that their frameworks arise from and, whether knowingly or unconsciously, support the society they wish to criticise. He argues that the values of a radical environmentalism cannot simply be incorporated in a moral framework which is derived from and supports our destructive society.

Although, in many respects, the papers in this section are very different, certain themes recur. One issue is what is meant by 'the environment'. Alternative definitions are discussed explicitly in the paper by Clift et al. Drawing on thermodynamic analysis, they distinguish between the 'system', ie the part of the universe to be studied, and the 'environment', defined by exclusion as anything which is not the system. However, they demonstrate that even applying this systems model to an (apparently) simple problem like the assessment of wind turbines as a method of generating power is not straightforward. The traditional engineering approach would be to view the turbine as the system and to attempt to assess its effects on the natural environment. Alternatively, the turbine can be seen as the system and the environment extended to include the natural and social environments. Or the local community can be viewed as the system and the turbine seen as a change to the surrounding 'natural' environment.

These examples raise the issue of whether we are primarily concerned about the impact which 'nature' (and human interventions in the natural world) has on us (ie the effect of the turbine on the community), or concerned about the effect that we as human beings have on nature (ie the effects that our use of wind turbines has on the non-human environment). Redclift discusses the development of science and social science in the nineteenth century and the way this has predisposed us to seek technical solutions to human-induced problems (which are perceived as problems because of their effect on human beings). Environmental management, in this analysis, is about mastering nature for the benefit of human beings. Smith argues that while 'experts' on environmental ethics operating within the accepted paradigms of moral theory tend to view the non-human world as of only instrumental value, one should recognise that many people do have genuine moral concerns for the non-human environment and a radical environmentalism is not anthropocentric.

A further issue in defining the 'environment' is whether to focus on the local or the global. Clift et al cite Cooper's (1992) distinction between The Environment, or the global environment and an environment, which brings with it the connotation of closeness, familiarity and what we know intimately. As Clift et al point out in relation to wind turbines, the same intervention may be perceived to have different impacts on the local and the global environments. Wind farms may be thought to have negative consequences for the local environment (in terms of changes to the landscape and noise pollution) but positive consequences for the global environment (in terms of sustainable energy). Redclift's chapter focuses on The Environment and argues that our 'ownership' of global environmental issues is affected by the ways in which we are removed from the consequences of our actions. Industrialisation in the countries of the North leaves 'ecological' footprints in countries thousands of miles away. But the consequences of actions in the industrialised countries are unpredictable, may be years in the future and bear little relationship to everyday decision making. In comparison, responses of local communities to local environmental problems are in tune rather than in conflict with consumerist values which incorporate a commitment to 'improve' one's lifestyle, so that a concern about *an* environment does not necessarily convert into a concern for *The* Environment. Parker's paper, on the other hand, focuses on ways of developing 'communities of environmental care' by devolving decision making about environmental issues to local communities and building on the particular concern that people feel for the place where they live. She sees this as a first step to developing appropriate ways of responding to global environmental issues, arguing that 'where sustainability is concerned we'd better all start doing it where we live'.

A second theme which runs through the papers is that of the role of the 'expert' in providing a valuation of the environment. There are a wide range of experts who may claim to be able to inform the public debate: pure scientists, engineers, economists, sociologists, psychologists, geographers, moral philosophers etc. Indeed, one of the functions of a book such as this is to provide a forum in which the experts can lay out their wares and show what their discipline can contribute to our understanding of the value of the environment. But how do laypeople make use of experts when it is increasingly clear that the 'answers' they produce are complicated and contradictory? Is the role of the expert to sell the 'right solution' to the public, given that, as Parker discusses, people cannot necessarily be trusted to come to a conclusion based on the 'right values' even if the 'right' decision-making processes are invoked? Or should the 'expert' be used as a resource, as someone who is skilled in the tools of a particular form of analysis, but who will provide a partial answer at best? To the extent that experts are inevitably prisoners of the society in which they live, as discussed in Smith's paper – trained to work within particular theoretical paradigms and constrained by the codes of their profession, however critical they or we may be of the dominant ways of valuing the environment – their role in society helps to support those values which they seek to criticise.

REFERENCE

Cooper, D. (1992) 'The idea of environment' in D. Cooper and J. Palmer (eds) *The Environment in Question*, pp 165–80, Routledge, London.

1 Values and Global Environmental Change

MICHAEL R. REDCLIFT
Environmental Section, Wye College

INTRODUCTION

Although it is usually conceded that values play a large part in the way we approach the environment, particularly the environment on our doorstep, the same concession is rarely made for the *global* environment. Global environmental change is often identified with physical processes 'out there', such as ozone depletion, biodiversity losses and, particularly, global warming. The global environmental agenda has, to some extent, been established by the natural sciences, working within a positivist tradition (Newby 1993). The reports of the Intergovernmental Panel on Climate Change (IPCC) are a case in point. The authority of the IPCC's deliberations stems, to some extent, from its 'scientific' objectivity, which influenced people such as the former British Prime Minister, Mrs Thatcher, into lending it their support (Boehmer-Christiansen 1993).

This chapter examines whether global environmental change (a phrase, significantly, that is often written with initial capitals) is as free from value considerations as many people believe, or hope. It goes on to explore three clusters of issues which suggest that an alternative approach needs to be taken.

Global environmental change can be understood in terms of three sets of issues, each of which forces us to examine our part in its construction: human relations with 'Nature', the need to live with increased uncertainty, and the extent to which our management of the environment reflects essentially human, rather than environmental, concerns. Each of these issues influences not merely the way we understand environmental problems, but also the way in which we can act to change them. They are also represented in the three major policy initiatives to have developed out of the Earth Summit in Rio de Janeiro in 1992: the Framework Climate Convention, the Biodiversity Convention, and the institutions responsible for establishing more sustainable practices at the international level (particularly the Commission for Sustainable Development and the Global Environment Facility) (Grubb 1993). In the final part of this chapter their relevance to this global agenda is considered.

Values and the Environment: A Social Science Perspective, Edited by Yvonne Guerrier, Nicholas Alexander, Jonathan Chase and Martin O'Brien. © 1995 John Wiley & Sons Ltd.

HUMAN RELATIONS AND 'NATURE'

The nineteenth century was a period in which the physical sciences saw spectacular progress, and most of the scientific disciplines assumed the identity they possess today. It was also a period, in Europe and North America, of enormous economic growth, and with economic progress came confidence. Looking back from the vantage point of the end of the twentieth century, this belief in progress, and the confidence that went with it, are the hallmarks of modernism (Redclift 1993).

Relatively rapid industrialisation, and the growth of towns, were 'global' phenomena because they served to incorporate other economic systems and other cultures. Globalisation in the latter part of the twentieth century has served to underline these links, changing the international economic division of labour, using technology and communications to provide global images, as well as markets, and seeking to preserve the exotic and unfamiliar ('the other'), whether through tourism or environmental campaigning, as items of consumption (Featherstone 1990; King 1991).

During the late nineteenth and early twentieth centuries, the opposition between nature and culture made room for the social sciences as autonomous disciplines; they grew up in the interstices between the ethical concerns of the humanities and the positivism of the 'hard' sciences. The insistence that human cultures were distinctive brought into question both the 'external' environmental determinism of some of the new sciences and the 'naturalism' of others, which saw human behaviour as the outcome of 'internal' biological forces, equally beyond our control (Benton and Redclift 1994).

Both of the imperatives provided by nature, the external environment and the human biological condition, were found wanting. It is not an accident that many of the issues which proved (and still prove) difficult for the social sciences to confront, such as eugenics, racism and the measurement of intelligence, lie at the crossroads of biology and social conditioning. In this sense the nature/culture dichotomy was both the springboard for the social sciences' advance, and the irresoluble problem they confronted (Benton and Redclift 1994).

Within the social sciences the discipline which benefited most from its identification with human purposes in the nineteenth and twentieth centuries was economics. Neoclassical economics grew out of the increasing confidence of industrial capitalism with its own success, and its refinements were linked to the problems faced by twentieth-century industrial economies (welfare economics, Keynesianism, development economics). Many of the issues which confront us as we approach the twenty-first century – the relationship between the production of goods and services and the satisfaction of our needs, as well as the social and environmental consequences of their production – elude mainstream economists. Many of the underlying assumptions which influenced economic reasoning, such as the effects of scarcity, now appear much less important than issues like the environmental consequences of economic behaviour, which played little part (Yearley 1991).

In the view of many neoclassical economists the significance of scarcity could be grasped through concepts such as the economic costs of resource acquisition. Pollution and the proliferation of so-called 'externalities' can be seen as manifestations of profligacy, rather than scarcity, and our inability to manage its consequences. As our dependence on economic techniques increases, the need for more inclusive systems of thought appears more urgent. We are forced back, inevitably, to consider our relations with nature, from which resources are derived.

Our increasing knowledge of biological systems has not enabled us to utilise them sustainably, and this is due in some part to the divorce which was effected in the nineteenth century between our understanding of the laws of nature and those of 'man'. We are faced by an interesting paradox. On the one hand the degradation of nature has called into question some of the values which contributed to the Promethean successes of the past. The rights of non-human species, and the primary obligations which we have to nature, are now regarded as politically important, not merely by deep ecologists. At the same time, many of those who espouse environmental concerns refuse to acknowledge that it is the way in which human societies are organised, and structured, which determines environmental problems.

What are the values generated from the management of the environment today? They clearly reflect the interface between society and nature, and the difficulty we experience in dealing with this interface. Environmental management itself suggests a mastery of nature and an ability to control the environmental consequences of our behaviour. The growing importance of scientific knowledge, and 'rationality' as the coda for this knowledge, together with our institutionalised behaviour and social commitments, have served to increase the appeal of technical solutions to human-induced problems. To provide solutions to environmental problems, however, we need look no further than the human societies which produce them; something which we seldom do (Beck 1992).

LIVING WITH UNCERTAINTY: THE IMPORTANCE OF TIME AND SPACE

Another consequence of the growing confidence of science has been the expectation of certainty. With the development of scientific techniques and methods the status of scientific prediction rose, and with it the status of scientists. Predicting environmental consequences has proved to be difficult, however, partly because of the complexity of environmental systems, and partly because of the unpredictability of human actions. Science is apparently successful in offering predictions which reduce uncertainty. However, science also collapses time and space, and increases the flow of knowledge and information available. This, in turn, tends to increase uncertainty and to fuel speculation about the basis on which decisions have been taken. We need to give close attention to the factors which have buttressed the claims of science to reduce uncertainty (Brown 1989).

First, many environmental problems involve high levels of human anxiety, associated with risks to human health, which appear to increase with the expansion of our knowledge. Secondly, since environmental science is an essential part of the solution to environmental problems, it follows that improved regulation, and greater technical expertise in addressing environmental problems, also serve to demonstrate the shortcomings in the application of science (Read 1994).

Global environmental problems in particular, such as ozone depletion and global warming, are not only complex in terms of their chemistry or biology, they are also apparently inaccessible to technical 'fixes'. Unlike the administration of antibiotics, or the inoculation of patients against the risk of contracting life-threatening diseases, changes in behaviour induced by environmental awareness, such as the purchase of aerosols free from CFCs (chlorofluorocarbons) or the use of lead-free petrol, do not ensure environmental safety. We know more but we are able to do less.

In addition, there is evidence from the growth of campaigning groups around environmental issues that the gap between 'lay' perceptions of the environment and 'expert' opinion is actually widening (Yearley 1991). Faced with a barrage of increasingly complicated and contradictory information about environmental risks, the layperson is likely to question the authority of science, and the confidence which politicians place in scientists. It soon becomes clear that the 'critical thresholds' which are endorsed by political leaders and expert witnesses, are themselves political compromises, framed to manage public apprehension. The more that the official environmental discourse may seek to dampen public apprehensions, the more it becomes clear that 'certainty' does not prevail.

Public anxiety is only part of the picture. If the environment exists in the specialist knowledge which we possess about it, there is less for the 'non-experts' to regard as their area of competence. This affects the 'ownership' of environmental issues. Research from developing countries has shown that the growth of specialist knowledge is related to the growth of non-specialist 'ignorance', and this observation is equally appropriate in the North. Doubts about the degrees of certainty associated with formal scientific knowledge are matched by alternative, holistic models of human relations with nature, which interpret 'facts' differently and which seek new ways of understanding, rather than an enlarged databank of information. It is clear that different values are held by different groups of people. Some groups, at least, are using the opportunity presented by scientific uncertainty to re-evaluate their values (Thompson, Warburton and Hatley 1986).

The two dimensions of uncertainty which deserve particular attention are the spatial and the temporal. We are accustomed to making most decisions on the basis of present time, and any future consequences play a smaller part in our calculations than do immediate consequences. However, environmental choices often bear little relationship to the decision making and dislocation of everyday life. They require an imaginative leap into the future, to the next generation or subsequent generations. The timescale of ecological processes, particularly those operating at the global level, makes it imperative that we attach weight to the future, and that what economists call the rates of discount reflect this importance.

Many environmental changes are also 'systemic' in the sense that they can only really be understood through the way in which systems change. Biodiversity is a case in point, since threats to individual species carry serious implications for ecosystems as a whole. The loss of one plant variety from a local ecosystem can jeopardise the survival of animal populations which are dependent on it. Since the timescale of ecological processes bears so little relationship to everyday decision making, it is important that we attach value to the loss of flexibility and variety in future environments.

The spatial dimensions of the environment are also important in any consideration of values. The environmental consequences of human activity are often experienced at several removes, not only in time but in space. The economic development of the industrialised countries, their diets and lifestyles, have been responsible for transforming the environments of developing countries located thousands of miles away. The 'ecological footprints' left by industrialisation and consumer wants in the North are not easily erased. This serves as a reminder that, while in the North we tend to regard the protection of nature as a fundamental ingredient of environmentalism, in the developing countries environmental issues often present themselves in terms of protection *from* nature. Perhaps we need to consider whether the driving forces behind global environmental change, including industrial growth and consumerism, increase the environmental security of people in the South or seriously threaten it?

The values generated in our society carry implications for the environment which are only dimly perceived most of the time. Consumerism implies a commitment to aspiration, to 'improve' one's lifestyle. A desire to own the fruits of technoscience is apparently within everyone's grasp. At the same time we are concerned should the environmental costs of progress arrive on our doorstep. The response of local communities to environmental problems – 'Not In My Back Yard' (NIMBY) – is a product of contemporary lifestyles in the industrialised countries, every bit as much as is concern about protecting the whale or tropical forests.

The process through which we are removed from the consequences of our actions, sometimes called 'distanciation', is illustrated in a number of ways. Among them is the way the enhanced greenhouse effect, through its impact on climate, is likely to increase perturbations in weather conditions, especially in the tropics, with increased occurrence of freak storms, drought and sea-level rise. The measures necessary to avert these risks are not difficult to determine, but the political will to act confronts widespread apathy and indifference.

ECONOMIC VALUES AND ENVIRONMENTAL MANAGEMENT

Neoclassical economics developed through making a number of assumptions about the environment. Natural resources such as water, soil and clean air were often depicted as 'free goods', meaning that they were available freely; they did not involve a charge. However, it is clear that environmental 'goods' are qualitatively different, in significant ways, from goods for which we do pay a charge. Clean water and air, unpolluted soils, are not available 'freely' in nature once human beings have had a hand in economic development. Environmental economics has been forced to consider the costs of cleaning up the environment, and of conserving natural resources to ensure their supply (Winpenny 1991).

Ecological economics is also concerned with wider questions which have eluded most economists since the nineteenth century. Attempts are being made to distinguish between 'wants' and 'needs', and between the way our needs are satisfied, for example through more consumer goods, and the needs themselves. The conditions under which goods and services are produced is a key question. At the same time the social and ecological *consequences* of their production are of concern to Green economists. Many argue that we should develop methodologies for arriving at 'utilisation values', that is, the value of goods and services throughout their lifetime. Such values would include the cost of waste disposal, the benefits from re-use or recycling, and the pollution or resource degradation associated with the use of raw materials in their manufacture.

Within environmental economics there are broadly three camps. The first camp argues that there is nothing to prevent us from placing economic value on the environment. Using prices and market instruments we can assign the real costs of environmental degradation. What is required is further refinement of methodologies such as contingent valuation, which enable us to approximate individual preferences for environmental goods and services. In the view of these economists the 'logic' of economic rationality can be used to manage the 'randomness' of nature (Pearce 1993).

A second camp takes a very different view. They argue that we cannot place a value on the environment, like that for human-made goods. Natural capital, in their view, is qualitatively different from human-made capital and should be treated as qualitatively different. Following Oscar Wilde's famous aphorism, we are in danger of knowing 'the price of everything and the value of nothing'. In the view of radical ecologists the logic of nature cannot be geared to the randomness of the market. As human beings we are part of nature, and cannot subject nature to our laws as we are subjected to natural laws (Ekins and Max-Neef 1992).

Between these apparently irreconcilable positions are others which probably attract considerably more support than is immediately evident. Some institutional economists such as Jacobs argue that we can, and should, develop economic methodologies which, in effect, 'value' nature (Jacobs 1991). However, we should also recognise that neoclassical economics is itself a social construction, and its development reflects the preoccupations of industrial capitalism. We can develop methodological tools which place more, or less, emphasis on the importance of market forces. If we wish we can propose guidelines, indicators for 'sustainability planning', which allow radical shifts in economic policy and thinking.

Unlike some radical political ecologists, people in this third camp propose that we intervene and regulate the environment, essentially to meet human purposes rather than follow imperatives in nature itself. They also agree that we will all be the richer if we examine the underlying social commitments which govern our lives, the maintenance of our present 'lifestyles' and patterns of consumption. However, unlike deep ecologists for example, mainstream environmental economists believe that changes in human behaviour can be induced through policy instruments and interventions.

It is also important to distinguish between analytical positions such as those found within environmental economics or the sociology of science and the value commitments of a society. To some extent analytical positions can play the part of, or even displace, other systems of values. We have only to reflect on the central role which neoliberal economics has attributed to the 'choices' of individuals in the marketplace, or what Huber has called 'ecological modernisation' through which business has sought to incorporate environmental costs in its range of products and services (Mol and Spaargaren 1990).

These are examples of the close relationship between the values of the wider society and those which govern environmental questions. It would be surprising if core values such as 'individualism', 'private property', 'choice' and 'independence', the political values which govern everyday actions and desires, were unrelated to the way in which we interact with our environment. However, it is much more difficult to specify the nature of this interaction, the variables at work and the lines of causation.

These positions themselves reflect a modernist discourse which still sees the human subject as universal and all knowing (Redclift 1993). They do not address the fallibility of human beings, most notably in our inability to reflect on the increased knowledge which we possess about the wider universe. If science is continually widening the frontiers of what we know, it is also revealing the extent of what we do not know. We are, in fact, seeking to interpret what we do not know in terms of what we know. At the very least this is a hazardous procedure.

GLOBAL ENVIRONMENTAL AGREEMENTS AND HUMAN VALUES

The international agreements which were signed at the Earth Summit in 1992 give expression to environmental values, many of them widely shared. At the same time these agreements, if they are to succeed in changing the way we manage our resources globally, require that we pay more than lip-service to the values we espouse. The institutional apparatus established at Rio de Janeiro, as much as the agreements themselves, provides evidence of the difficulty in providing a consensus for global environmental management (Thomas 1993).

The Convention on Biological Diversity provides a useful example of the limits of international agreements which are not underwritten by shared values. This Convention seeks to preserve biological diversity on the planet through protecting both species and ecosystems. It also established policy options for making use of these biological resources and associated technologies (Grubb 1993). Like the Climate Change Convention it lacks specific commitments that are binding on the parties. However, it is also broader and more self-contained than the agreement on climate.

During the negotiations the developed countries sought access to, and protection for, the developing countries' biological resources. The developing countries, for their part, sought access to the biotechnologies of the North on preferential terms. Perhaps the principal value to be affirmed in the Convention is that nation states have 'sovereign rights' over the biological resources within their territories, and that the benefits which flow from these resources should be shared more equitably and on 'mutually agreed terms'. The Convention endorses the idea that individual countries should develop plans for protecting biodiversity, which will be submitted for scrutiny.

As Tickell has observed, the Biodiversity Convention addresses issues which are fundamentally ethical in nature:

> There is an increasing awareness, especially in societies that have done most to destroy other forms of life, that humans have some kind of ethical responsibility for the welfare, or at least the continued existence, of our only known living companions in the universe. (Tickell 1994, p2)

However, action to arrest the destruction of biological systems, in the interest of the entire globe, has floundered on vested interests. The countries of the South, not surprisingly, insist on their sovereign rights to manage their own ecosystems. Those of the North, notably the United States, have equally insisted on their 'rights' to have access to these resources, on the grounds that to prevent their exploitation would fly in the face of scientific progress. The modernist conviction that 'progress' can be arrived at through the development of science prompted, of course, by market forces lies at the heart of the value system of the industrial countries. The United States government, under President Bush, even refused to sign the Convention.

Tickell has vividly demonstrated how we are dependent on biological resources, how they underwrite our daily lives:

> At present we take as cost free a broadly regular climatic system with ecosystems, terrestrial and marine, to match. We rely on forests and

vegetation to produce soil, to hold it together and to regulate water supplies by preserving catchment basins, recharging groundwater and buffering extreme conditions. We rely on soils to be fertile and to absorb and break down pollutants. We rely on coral reefs and mangrove forests as spawning grounds for fish and wetlands, and on deltas as shock absorbers for floods . . . There is no conceivable substitute for these natural services. Yet we cannot continue to assume that this natural bounty will continue. (Tickell 1994, p3)

It is clear that values are implicit in what we take for granted from natural systems, as well as what we propose to do to protect these systems. At the same time, the process of economic development enshrines a different set of values. The Brundtland Commission, which reported in 1987, sought to enlarge this debate and to make our value preferences more explicit (Brundtland 1987). Unlike the reports of the IPCC it did not purport to be a value-free document, but freely admitted to political objectives, many of which were subsequently incorporated in Agenda Twenty One. The idea of 'sustainable development' as a way of informing policy cannot be divorced from the attempt to integrate quite different systems of values.

Much of the confusion accompanying the discussion of sustainable development, and the drawing up of international agreements, stems from the relationship between our values and our knowledge about global environmental problems. The scientific controversy accompanying global climate change, and the deliberations of the IPPC, have suggested that increasing our knowledge about future climate change and its impacts will enable us to adopt more appropriate values, emphasising long-run sustainability over short-run economic gain. However, the evidence for this assumption is weak. Rather, it might be asserted that until we address the environmental problems associated with our current values, there is little likelihood that we will be able to make much use of the knowledge which is accumulating about the global environment. As Tickell argues:

our ignorance of species and ecosystems is profound, not only of present ecosystems and species, but of their future uses and services. It is an understatement to refer to this level of ignorance as mere uncertainty. (Tickell 1994, p4)

The major provisions of the Framework Convention on Climate Change mark an important watershed in international agreements to protect the environment. The Convention established the principle that action to start addressing the problems of climate change should not wait on the full resolution of scientific uncertainties. It also asserts that developed countries should take the lead in introducing measures to reduce the threat of global warming. Finally, it endorses the idea that developed countries should compensate the developing countries for any additional costs which they might incur in taking measures under the Convention.

Superficially, at least, the goal of sustainable development is one publicly espoused by most governments. Most of the governments in the North have signed, and in some cases ratified, agreements which endorse a set of principles and values which place global sustainability above vested interests and short-term economic advantage. However, at a more profound level there is little agreement about the 'values' which need to inform sustainable development. The 'natural services' provided by the environment are acknowledged, but the assumption that they will continue to be provided is still made. Real environmental costs and benefits are scarcely acknowledged in the day-to-day economic management which determines their use.

Similarly, rather than using the precautionary principle to help provide for more flexible responses to uncertainty, most policy is still formulated against unsustainable assumptions, about population, military expenditure and economic growth. Global inequalities, particularly between North and South, are part of the 'taken-for-granted' assumptions behind international agreements in 'non-environmental' areas such as the liberalisation of trade. Inequalities within developing countries, we are regularly told, are part of the price that such countries pay for the absence of 'development'. However, evidence that economic growth has particularly adverse effects on the newly industrialising countries' environments should lead us to question whether successes in market economies really are a prerequisite for better environmental management in these countries.

Until sustainable development informs our behaviour at all levels it is unlikely that global agreements to manage the environment will do anything but scratch the surface of the problem. Environmental values, including the way we relate to nature and future uncertainties, need to drive changes in our underlying behaviour and the assumptions about the environment which form part of routinised practice. This observation, of course, is itself a value judgement. We cannot expect much success at sustainable development until we acknowledge that both science and politics are based on values which are unsustainable. The challenge is to provide a confident vision of a society far removed from the one we know: we require a vision that is nothing short of an alternative future.

REFERENCES

Beck, U. (1992) *Risk Society*, Sage, London.
Benton, T. and Redclift, M. R. (1994) 'Introduction' in T. Benton and M. R. Redclift (eds) *Social Theory and the Global Environment*, Routledge, London.
Boehmer-Christiansen, S. (1993) 'Scientific consensus and climate change: the codification of a global research agenda', *Energy and Environment* **4**,4, 163–4.
Brown, J. (ed) (1989) *Environmental Threats*, Belhaven Press, London.
Brundtland Commission (World Commission on Environment and Development) (1987) *Our Common Future*, Oxford University Press, Oxford.
Ekins, P. and Max-Neef, M. (eds) (1992) *Real-Life Economics*, Routledge, London.
Featherstone, M. (ed) (1990) *Global Culture*, Sage, London.

Grubb, M. (1993) *The Earth Summit Agreements: a Guide and Assessment*, Earthscan/Royal Institute of International Affairs, London.

Jacobs, M. (1991) *The Green Economy*, Pluto Press, London.

King, A. (ed) (1991) *Culture, Globalization and the World System*, Macmillan, London.

Mol, A. P. J. and Spaargaren, G. (1990) 'Sociology, Environment and Modernity', Paper presented to International Sociological Association Conference, Madrid.

Newby, H. (1993) *Global Environmental Change and the Social Sciences: Retrospect and Prospect*, Economic and Social Research Council, Swindon.

Pearce, David (1993) *Blueprint Three: Measuring Sustainable Development*, Earthscan, London.

Read, P. (1994) *Responding to Global Warming*, Zed Books, London.

Redclift, M. R. (1993) 'Sustainable Development: Needs, Values, Rights', *Environmental Values* **2**,1, Spring, 3–20.

Thomas, C. (ed) (1993) 'Rio: Unravelling the Consequences', *Environmental Politics* (Special Issue) **2**,4, Winter, 1–27.

Thompson, M., Warburton, M. and Hatley, T. (1986) *Uncertainty on a Himalayan Scale*, Milton Ash Edition, London.

Tickell, C. (1994) 'Socio-political Perspectives on Biodiversity', Working paper, Green College, Oxford.

Winpenny, J. T. (1991) *Values for the Environment*, HMSO, London.

Yearley, S. (1991) *The Green Case*, HarperCollins, London.

2 Environmental Perspectives and Environmental Assessment

ROLAND CLIFT, KATE BURNINGHAM AND RAGNAR
E. LÖFSTEDT
Centre for Environmental Strategy, University of Surrey

INTRODUCTION

The three authors of this chapter come from different academic backgrounds – respectively engineering, sociology and geography. Behind the obvious differences in professional vocabulary between these disciplines lie differences both in the way the environment is viewed and in the aims of analyses of environmental impact. This contribution explores the nature of some of these differences, and goes on to identify some of the benefits to be gained from attempting to synthesise our different approaches. We base the paper around environmental assessment of a specific issue – wind turbines to generate electric power. The methodology of one approach to environmental assessment is summarised, to show how it is built on a specific but restrictive definition of 'The Environment'. We then examine ways in which our diverse perspectives and professional backgrounds might be combined to move towards more genuinely holistic and acceptable approaches to environmental assessment.

Wind power is an issue of current debate. It was originally hailed as an important component of a sustainable society based on renewable energy (and we deliberately refrain here from attempting to define 'sustainable' and 'renewable'). More recently, some of the problems of wind farming have been invoked to question the desirability of this form of renewable energy. In this chapter, we will be less concerned with economic questions than with 'environmental' issues (Halliday and Jenkins 1994): the environmental impacts of making and demolishing the turbine, local environmental effects (primarily aesthetic intrusion on the landscape, noise, bird-kills, and interference with television reception) and risks (eg detachment of a blade from a moving turbine).

ENVIRONMENTAL SYSTEM ANALYSIS

The specific approach to environmental assessment discussed here is life cycle assessment (LCA), alternatively known as cradle-to-grave analysis. LCA has

Values and the Environment: A Social Science Perspective, Edited by Yvonne Guerrier, Nicholas Alexander, Jonathan Chase and Martin O'Brien. © 1995 John Wiley & Sons Ltd.

become established as a widely used environmental management tool. In the public sector, it provides the basis for Ecolabel schemes, intended to identify consumer products which genuinely have excellent environmental performance (see Clift 1994). A code of practice for the technique has been developed by the Society of Environmental Toxicology and Chemistry (SETAC 1993). The account here follows the simpler sequence outlined by Udo de Haes et al (1994).

'SYSTEM' AND 'ENVIRONMENT': THE THERMODYNAMIC BASIS

It is worth recalling that the word 'environment' in the *Oxford English Dictionary* definition of 'physical surroundings and conditions' entered the English language via the thermodynamic literature as a translation or transliteration of the French *environnement*. The thermodynamic concept is illustrated in Figure 2.1. In any thermodynamic analysis the first step is to define the 'system', ie the part of the universe which is to be studied (for example a steam engine or a kilogramme of steam). The environment then constitutes the 'physical surroundings and conditions' outside the system, ie the environment is defined by exclusion. Some social scientists are liable to find this stark, superficially arbitrary delineation between 'system' and 'environment' unpalatable ('. . . but who says you can draw the boundary there?'). However, any form of quantitative system analysis must start by defining the system to be analysed.

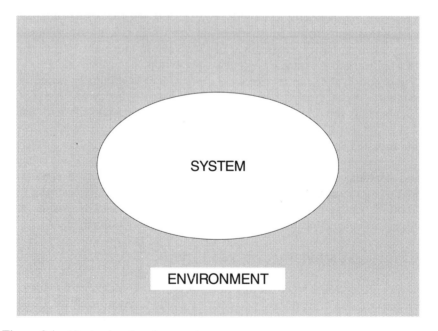

Figure 2.1 'System' and environment

Definition of the 'environment' as 'that which is not the system' spreads from thermodynamics to all branches of science and engineering which draw on thermodynamics. Albert Einstein's famous dictum 'the environment is everything which is not me' can be seen as an example. The distinction between 'system' and 'environment' is at the heart of environmental management and assessment (see Figure 2.2). The 'system' is now the human activity which provides goods and services as its output. To provide this output, the system requires inputs of materials and energy, ie it consumes renewable and non-renewable resources, which pass from the environment into the system. To provide the desired outputs, the system also generates unwanted outputs – emissions and solid wastes – which pass from the system to the environment as environmental pollution. Environmental system analysis is concerned with these flows of material and energy between the environment and the system.

The kind of diagram illustrated by Figure 2.2 is part of the normal vocabulary of system analysis, whether applied to environmental or other problems.

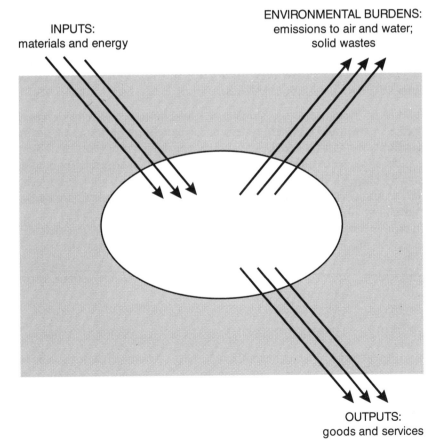

Figure 2.2   Environmental system analysis (after Azapagic and Clift 1994)

Environmental impact assessment deals with identified manufacturing sites; therefore the 'system' is the factory or process, or, in the specific example here, the wind turbine installed and operating. Environmental life cycle assessment deals with the whole system needed to provide a function or service, from 'cradle' (primary raw materials and energy) to 'grave' (final waste and residue). Therefore the 'system' in LCA is the cradle-to-grave flow of materials and energy needed to provide the output – in this case, power generated by the turbine (see, for example, Clift and Longley 1994). LCA starts from the mineral deposits extracted to make the turbine, considers the environmental impacts of mining the ore, processing it into refined metal, fabricating the machinery, transport and erection, demolition at the end of the service life, and then re-use or final disposal of the material.

Defining the boundary of the system – and hence, by exclusion, defining 'The Environment' – is formalised in LCA in the step known as 'goal definition'. Clearly this step is critical. LCA has, with some justification, been criticised because defining the system boundary can determine the outcome of, for example, comparison between two alternative products. There is therefore a strong argument for subjecting goal definition to some form of public scrutiny, and we return to this point below.

QUANTIFYING THE EXCHANGES: THE LIFE CYCLE INVENTORY

The kind of system analysis summarised by Figure 2.2 is only designed to describe interactions between system and environment in terms of *fluxes*, ie quantifiable flows of materials and energy. For the wind turbine, the inventory compiles all the emissions and wastes produced and other resources consumed – notably energy – in mining, processing, fabrication etc. The data on resource inputs to and emissions and residues from the productive system are known as the inventory table.

It is immediately clear that conventional LCA includes only the quantifiable impacts of making and demolishing the turbine. Noise, bird-kills, television interference and aesthetic intrusion are all in different categories. These other environmental effects would be represented as 'non-flux interactions' between system and environment.

QUANTIFYING ENVIRONMENTAL EFFECTS: CLASSIFICATION

At its present stage of development, LCA treats environmental impacts as independent of the place where an emission arises; thus a tonne of sulphur oxide is simply a tonne of sulphur oxide, no matter where it is emitted. This is clearly unsatisfactory, even as a proposition in environmental science; the effect of the tonne of sulphur dioxide depends on where it is deposited, which depends in turn on the place and atmospheric conditions where it is emitted into the atmosphere. Localised environmental impacts are clearly significant in our example:

the noise and aesthetic intrusion of a wind turbine cannot be addressed on the 'globalised' level to which LCA aspires.

Conventionally, LCA proceeds by converting the fluxes in the inventory table, typically covering hundreds of substances, into a much smaller number of quantified environmental impacts or 'effect scores'. For example, all atmospheric emissions which could contribute to depletion of stratospheric ozone are weighted according to their 'ozone depletion potential' to give a total aggregated value for this one environmental effect (see Guinée et al 1993). For emissions which contribute to genuinely global environmental impacts, this approach has theoretical justification, although the numerical values for parameters such as 'ozone depletion potential' may be highly uncertain. One approach to classifying localised environmental impacts (see Guinée et al 1993) is based on the 'critical volumes' approach. Any specific emission is taken to have a threshold concentration below which the local ecosystem can tolerate it without damage. The 'critical volume' of air or water needed to dilute the substance to this tolerable concentration is then taken as a measure of the ecotoxicological significance of the emission. Again, the critical volume is taken to be independent of location. Thus LCA treats both global and local environmental effects as mechanistic, deterministic and quantifiable.

COMPARING ENVIRONMENTAL EFFECTS: VALUATION

Ideally, the objective of LCA is to reduce these effect scores to a single numerical value which represents the environmental impact of providing the functions which the system delivers. At the most reductionist level, this requires quantitative relative values to be attached to the different environmental effects. There are two issues here: how to fix comparable values to different effects, and how to deal with the possibility of these values differing between communities. If the relative values vary between different communities, then the LCA ideal of global valuation must be abandoned. This has various practical implications. One of these is that a uniform measure of environmental performance for a product is impossible. It then follows that an Ecolabel relevant throughout an area such as the European Union will be an impossibility (Udo de Haes et al 1994).

The wind turbine example illustrates the difficulties associated with the valuation of environmental impacts. From the viewpoint of an energy utility, the most important consideration will be energy consumed and generated. Therefore the life cycle energy consumption (ie the energy used in making, erecting and demolishing the turbine), sometimes called the 'embodied energy' will be of prime concern. The balance might be expanded to allow for the energy needed to make high-resolution TV aerials to overcome the interference of the turbine. If the embodied energy exceeds the energy yield from the machine during its foreseeable service life, then there is simply no energy benefit in building the turbine. Going beyond the LCA methodology, there is a conflict between the treatment of 'flux' and 'non-flux' interactions between system and environment. For example, ecological assessment of emissions in the manufacture of the turbine includes the toxicological effects of these emissions on bird populations, but ignores the more dramatic effect of the turbine on birds which happen to fly into it. Noise and aesthetic intrusion raise different questions. At this point, we have exceeded the scope of the thermodynamically based analysis summarised in Figure 2.2, and it is time to invoke a different perspective.

SOCIAL SCIENTIFIC OBJECTIONS TO THE 'SYSTEMS' APPROACH

THE SOCIAL ENVIRONMENT

As outlined above, from an engineering perspective the wind turbine is considered to be the 'system', with the 'environment' being everything else. This 'everything else' might include the social as well as the natural environment – see Figure 2.3 (although it would be unusual for engineers to use such a broad conception). A possible contribution for social scientists would then be to consider the positive and negative impacts which the system has on its social environment. However, it soon becomes clear that this is no straightforward task, and that it does not fit well into the system/environment division.

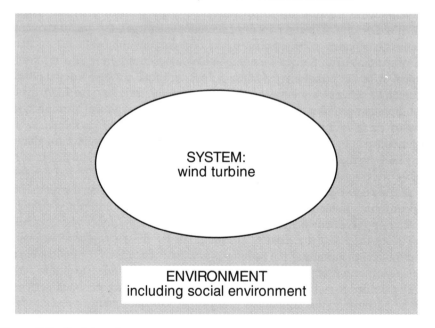

Figure 2.3   Social environment

Assuming for the moment that the existence of the system is taken as given, some decision has to be made about what is the 'social environment'. Is it just the local context (the area or community in which the turbine is sited) or does it include a wider view of society? Wind turbines are often considered to have unacceptable negative impacts on the local environment (in terms of changes to the landscape and noise pollution), but might be considered to have positive outcomes for society more widely through the provision of cheap and renewable energy. The systems approach can define 'cheap' and 'renewable', but not 'positive'.

The perspective of political ecology (an area of resource geography) draws attention to the need to consider the historical, cultural and community context of developments. This wider dimension is important not only in a consideration of impacts but also in terms of the factors which facilitate or constrain the development. For instance, in order to understand how a wind turbine comes to be built in a particular place at a particular time, attention needs to be paid to wider aspects of energy policy. Were it not for the Non Fossil Fuel Obligation (NFFO), wind turbines would not be built in the UK. This attention to the context in which wind turbines are developed does not fit easily into the engineering model which takes the existence of the system for granted.

Even if it is decided to include the impact of turbines on both the local area and on wider society, another problem is apparent – how to compare these very different levels of impact and decide whether overall wind farms have a positive or negative effect on the social environment. Not only are social impacts, such as the effect of noise or aesthetic changes, notoriously hard to quantify, but the analytic situation faced in evaluating the overall effect of a project often involves comparing essentially incomparable effects (see Wolf 1983).

An example of the first point is provided by disputes over the extent of the impact of noise. Although levels might be scientifically measured and declared reasonable, this assessment may well be challenged by local people who claim that the noise interferes with aspects of their everyday lives. For example, in a newspaper interview with a family living close to a wind farm the noise is described as 'not in itself deafening or intolerable . . . an everyday noise' but for the family this is said to render 'their house . . . worthless and their life intolerable' (*The Guardian* 11 March 1994). Thus alternative evaluations of the extent of an impact often exist. Whether noise is considered to be a problem will depend on a range of features including features of the local context and the sensitivity of individuals.

This goes beyond an acknowledgement that the same impact will have a different effect in different localities, and draws attention to the fact that even in the same place different individuals and groups will be affected differently by an impact. It follows that the objection to the aggregation of social impacts is not based only on the grounds of technical difficulty. If impacts are quantified and some overall measure of the social impact of wind turbines is derived, the way in which different social groups are affected is masked (Rohe 1982). The operation of the wind turbine may have unacceptable social consequences for some people while others benefit. This question of who is affected by a project and how is an essential one for social scientists, but cannot be incorporated readily into the flux or non-flux impacts of Figure 2.2.

The second objection is to the notion implicit in the engineering model that environmental impacts are caused in a mechanistic fashion. Although wind turbines obviously 'cause' noise and changes to the landscape, the identification of these as problems is not automatic. Some aspects of the wind turbine may be considered to be problematic while others are ignored (for instance, in some contexts noise will become an issue which leads to public concern and protest, while in others the issue may be the appearance of the site or perhaps nothing at all). In this sense the impacts of the wind turbine are socially constructed; local people are active in deciding what constitutes a problem for them and in deciding how it should be responded to (Burningham 1994).

As noted above, environmental system analysis tools such as LCA treat impacts as deterministic. At best, risks might be incorporated using probabilistic assessment tools, essentially predicting a number of human fatalities arising from the activity or project. However, this risk assessment approach ignores extensive research in the field of risk perception (eg Tversky and Kahneman 1974; Fischhoff et al 1978; Slovic 1987) which shows that the lay public defines 'risk' using a range of qualitative variables which differ considerably from those invoked in expert prediction of risk. This introduces a further problem in the systems approach, beyond that represented by different reactions to apparently quantifiable deterministic impacts such as noise. It is the public perception of risk which determines the acceptability of different technologies. For example, even if it were to be shown that more human fatalities would result from generating a certain amount of electrical power by wind turbines than by, say, nuclear generation, then this might still have little bearing on the relative acceptability of nuclear vs other sources (cf Löfstedt 1994). Thus the 'environment' cannot be treated simply as responding to stimulus from the 'system'.

THE IDEA OF SOCIAL SYSTEMS

An alternative way of conceptualising the environment in this example would be to retain the basic notion of a system surrounded by an environment, but to consider the local community as the system and the building of the wind turbine as a change in the surrounding 'natural' environment (see Figure 2.4). However, this model is no less problematic.

Social scientific systems theories treat society (or some aspect of it) as a system with a tendency to equilibrium, and analysis focuses on what role the various subsystems of society play in maintaining the efficient functioning of the whole. Societies are considered to be basically orderly stable entities which are held together by a consensus of values and priorities. As a structural theory, functionalism considers the ways in which individual and group behaviour is

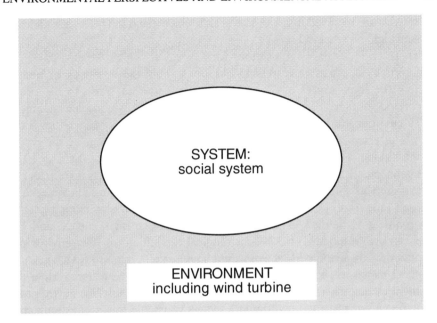

Figure 2.4   Social system

constrained or affected by the structures of society – the norms and values of the social system. However, critics of the approach have drawn attention to its weakness at explaining social change; the difficulty of determining whether a society is in a state of equilibrium; and the assumption that societies tend towards consensus rather than conflict (see for instance Cuff et al 1990).

With our wind turbine example we would have to consider whether the system of analysis consists of just the local community (and if so how the boundaries of that community will be defined) or whether it incorporates wider aspects of society. If the system of analysis was defined (for simplicity's sake) as the local community, we would have to decide how to deal with conflicts of views and interest within the area and changes over time. We would also have to consider whether it makes any sense to think of a local community in terms of a system in equilibrium and how to gauge what constitutes a significant change to that system.

A broader objection to structural theories is posed by action approaches (or interpretive sociology) which take issue with the idea that human action and interaction are simply determined by societal structures, arguing for the agency of human beings and their active role in interpreting and acting on the social world. From this perspective social life is treated not in terms of sets of structures, but in terms of sets of processes which members of society employ to make sense of the world. Rather than providing an overall view of society, interpretive sociologists investigate social situations from the point of view of those who are involved in them, and seek to understand the processes and concepts which they themselves use.

These critiques have implications for the way in which the environment is conceptualised. Apart from raising questions about the adequacy of treating society (or some aspect of it) as a system, attention is drawn to the central importance of meanings. The systems model is essentially a mechanistic model – the environment is treated as something which has direct effects on, and in turn is affected by, the system. Although it is useful to recognise the way in which human societies interact with their natural environment (Dunlap 1979), this conceptualisation of the environment/society relationship is insufficient. To treat society as something which relates to the environment in such a mechanistic way fails to do justice to the role of human actors in determining how the environment is defined, what counts as an environmental impact, and how impacts are responded to. A sociological definition of the environment informed by an action perspective would begin with questions of what the environment means to people and what effects that has on their actions, rather than seeing their actions as determined by, and in turn determining, the environment.

A focus on what the environment means to people raises questions about what environment is being discussed. Cooper (1992) draws a distinction between the notions of *The* Environment, or the global environment, the whole natural order; and *an* environment as something in which a creature (or person) is at home. His distinction is not simply one of scale (between global and local) but draws attention to the degree of closeness or familiarity which people have with the environment. He argues that *The* Environment is simply too big for people really to feel any association with, whereas *an* environment is what they are at home in, what they know intimately.

The wind turbine example illustrates nicely the use of different conceptions of the environment and the extent of disagreement about what is 'good' or 'bad' for it. In disputes over the siting of wind farms, both sides frame their arguments in environmental terms; those who are for the technology argue that it is good for the environment as it provides a clean and renewable energy source, while those who oppose it claim that it is bad for the environment on the grounds that it ruins beautiful landscapes and destroys the tranquillity of the countryside. In Cooper's terms those for the technology are talking about *The* Environment, while those who oppose it are talking about *an* environment.

Cooper's insistence that it is the environment which people know intimately which is of prime importance to them runs counter to the definition of the environment as 'everything which is not me'. Cooper writes:

> an environment is understood as a network with which a creature has a practical, smooth and unreflective familiarity, one it is 'at home' in. The creature is, in this way, part of its environment, though one could as truly say that the environment is part of it. (1992 p178)

In this definition people are not considered to be separate from the environment but are part of it; the environment is not conceptualised as an entity distinct from the meanings which they attribute to it. Interestingly this conceptualisation is akin to a secondary definition of the word 'environment' provided by the *OED* – 'conditions or circumstances of living'.

A definition of the environment which acknowledges the centrality of what an/the environment means to people has implications for the practice of assessing environmental impacts. First, it forces the recognition that the environment does not have a uniform meaning and consequently environmental change will not be responded to similarly, or have the same consequences, in all situations. A range of features of the local political and cultural situation will inform whether the condition of the environment is considered acceptable and what strategies are appropriate for maintaining or improving it. Although there is a need for basic standards, flexibility and adaptability to local circumstances are essential.

Finally, in the systems model all social responses and interpretations of the environment are relegated to 'non-flux' exchanges. The terminology suggests that these factors are not central to the way in which the system is affected by or responds to its environment. An approach which stresses the importance of understanding what the environment means to people inverts this assumption and suggests that social meanings are fundamental to the way in which the environment is experienced and affected.

PESSIMISM OR RESOLUTION?

The history of engineering projects and public inquiries illustrates the problems which arise when the different perspectives set out in this chapter are not brought together. At root, the scientific approach regards environmental assessment as a problem in establishing the best scientific predictions for the interactions between system and environment, and using these predictions as the basis for a rational decision. However, social scientists (and arguably the lay public) may well wish to see the planning process as a means of achieving some form of social consensus. Failures to reconcile these perspectives are obvious enough, and not just in the frequent public dissatisfaction over planning inquiries. It has been argued (Myers and White 1993) that many of the losses resulting from flooding of the Mississippi river in 1993 could have been avoided if the work of the natural hazards research community had been considered more carefully.

It is insufficient to state that engineers should include more social science in their work, or that social scientists should recognise the constraints within which engineers work. This easy conclusion ignores the fact that the aims of social scientific and engineering analyses often differ radically. LCA aims to provide data for use by decision makers. As a consequence it has to produce a 'result', ie an assessment. However, social scientific analyses usually aim to increase 'understanding', often without regard to decision makers. These varia-

tions in aim account for many of the differences in method and assumption outlined above. Furthermore, if aspects of social science are incorporated into
assessments, the expense and duration of the research can increase dramati-
cally. A totally comprehensive procedure for environmental assessment would
be complex, unwieldy and impractical. It is more realistic to regard environmental assessment methodologies (including life cycle assessment) as basic
tools only, to be adapted and shaped for specific applications. Even this rather
obvious conclusion runs entirely counter to the current development of firm
codes of practice for environmental assessment (eg SETAC 1993).

CONCLUSIONS

The concepts of 'system' and 'environment' will remain central to any public
debate over activities which have an impact on local communities and ecosystems. The fact that these concepts have completely different significance for
different groups of professionals adds a further dimension to the different popular constructions which usually emerge in public inquiries. By exploring the
differences in interpretation more openly, it may be possible to reconcile the
different values attached to 'The Environment', 'our environment', impacts,
perceptions of environmental risk etc. In the specific case of life cycle assessment, this requires goal definition to be exposed to some form of public
scrutiny. In fact, this has already been suggested as a way of improving the
credibility of the European Ecolabel scheme (Udo de Haes et al 1994). More
generally, the whole process of defining meanings and values needs to be
brought back into the public arena and made an integral part of policy debate
and specific enquiries.

The specific case of a wind turbine examined here has highlighted the
importance of the local context. To develop the methodology of life cycle
assessment, it will be essential to build in consideration of local impacts and
risks. More generally, this undermines the concepts of a universal best environmental option. It poses a real problem for multinational organisations which
may wish to apply common environmental standards wherever they operate, but
must then recognise that environmental sensitivies and social priorities will
differ between different places. Incorporating local values into environmental
assessment – reconciling 'The Environment' with 'our environment' – requires
a radical review of planning and policy which will only be productive if it
attempts consciously to reconcile different professional perspectives.

REFERENCES

Azapagic, A. and Clift, R. (1994) 'Allocation of Environmental Burdens of Whole-
System Modelling: The Use of Linear Programming', in G. Huppes (ed) Workshop
on *Allocation in LCA*, Society of Environmental Toxicology and Chemistry,
Brussels and Pensacola.

Burningham, K. (1994) 'The Social Construction of Social Impacts: Insights from a Case Study of the Social Impacts of a Road Scheme', paper presented at International Association for Impact Assessment annual conference, Quebec City, July.

Clift, R. (1994) 'Life Cycle Assessment and Ecolabelling', *Journal of Cleaner Production* 1, 155–9.

Clift, R. and Longley, A. J. (1994) 'Introduction to Clean Technology', in R. C. Kirkwood and A. J. Longley (eds) *Handbook of Clean Technology*, Chapter 6, Blackie, Glasgow.

Cooper, D. (1992) 'The idea of environment' in D. Cooper and J. Palmer (eds) *The Environment in Question*, pp165–80, Routledge, London.

Cuff, E-C., Sharrock, W. W. and Francis, D. W. (1990) *Perspectives in Sociology*, Unwin Hyman, London.

Dunlap, R. (1979) 'Environmental Sociology', *American Review of Sociology* 5, 243–73.

Fischhoff, B., Slovic, P., Lichtenstein, S., Read, S. and Combs, B. (1978) 'How Safe is Safe Enough? A Psychometric Study of Attitudes Towards Technological Risks and Benefits', *Policy Studies*, 9, 127–52.

Guinée, J. B., Heijungs, R., Udo de Haes, H. A. and Huppes, G. (1993) 'Quantitative Life Cycle Assessment of Products – 2: Classification, valuation and improvement analysis', *Journal of Cleaner Production*, 1, 81–92.

Halliday, J. and Jenkins, N. (1994) 'The Environmental Impact of Wind Energy', *Power Generation and the Environment*, pp237–52, Institution of Mechanical Engineers, London.

Löfstedt, R. E. (1994) 'The Barsebäck Power Plant: Fairness Across Borders', working paper, University of Surrey.

Myers, M. F. and White, G. F. (1993) 'The Challenge of the Mississippi Flood' *Environment*, 6, 9, 25–35.

Rohe, W. (1982) 'Social Impact Analysis and the Planning Process in the United States: A Review and Critique', *Town Planning Review*, 53, 367–82.

SETAC (1993) *Guidelines for Life-Cycle Assessment: A 'Code of Practice'*, Society of Environmental Toxicology and Chemistry, Brussels and Pensacola.

Slovic, P. (1987) 'Perceived risk', *Science*, 236, 280–85.

Tversky, A. and Kahneman, A. (1974) 'Judgment under uncertainty: heuristics and biases', *Science*, 185, 1124–31.

Udo de Haes, H. A., Bensahel, J. F., Clift, R., Fussler, C. R., Griesshammer, R. and Jensen, A. A. (1994) *Guidelines for the Application of Life-Cycle Assessment in the EU Ecolabelling Programme*, European Commission, DG XI-A-2, Brussels.

Wolf, C. P. (1983) 'Social Impact Assessment: A Methodological Overview' in K. Finsterbusch et al (eds) *Social Impact Assessment Methods,* pp15–33, Sage, Beverly Hills.

3 Enabling Morally Reflective Communities: Towards a Resolution of the Democratic Dilemma of Environmental Values in Policy

JENNETH PARKER
Centre for Continuing Education, University of Sussex

INTRODUCTION

The democratic focus of this chapter is prompted by the lack of debate around the social and political implications of proposed environmental reforms. I will argue that this is a matter of particular concern as, in my view, a revival of local democracy is a necessary condition for the development of sensitive environmental care. As a community education practitioner and as a philosopher I am worried about the negative consequences of top-down models of environmental value change. In particular, I believe that there has been inadequate analytical attention to questions of political legitimacy by some of those proposing new public policies based on environmental values. This chapter also attempts to clarify issues which are important in my own practice. My concerns are informed by feminist epistemological debates on knowledge and power (Garry and Pearsall 1989; Alcoff and Potter 1993) – in particular questions of the reflexivity of theory.

The democratic dilemma between the 'right values' and the 'right processes' arises in societies with democratic moral and political commitments. Problems of 'right values' and democratic processes have not been created *de novo* by environmental value considerations but have been central problems in the history of democratic theory. In *The Republic*, Plato deplored the fact that the mass of people could be persuaded of wrong values by the clever oratory of the Sophists. He drew the conclusion that democracy could not be trusted. The 'right values' of equality prompted by the enlightenment philosophers supported democratic processes (hereafter referred to as the 'right processes'). Political equality was supported by the supposedly equal 'rational capacity' of human beings (see Mill 1974).

The 'democratic dilemma' arises for those moral reformers in the democratic community who strongly hold certain values and still wish to uphold democratic

Values and the Environment: A Social Science Perspective, Edited by Yvonne Guerrier, Nicholas Alexander, Jonathan Chase and Martin O'Brien. © 1995 John Wiley & Sons Ltd.

values. The rational democratic route for individuals to take who want to influence value change is to engage members of the democratic community in debate. (There are also affective routes, for example through the arts.) However, moral reformers may find that one route towards encoding their values in law is via power, enabling them to exercise influence in a way which is fundamentally undemocratic. I will argue that questions of the integration of environmental values into public policy raise this dilemma in a particularly sharp form. It is my aim in this chapter to explore this problem and to suggest some strategies towards its resolution.

As my main example I will use a report detailing a system for the inclusion of environmental values in public policy. This report, which contains much of value, takes the 'right values' horn of the dilemma, consistent with the political practice which I describe as 'mitigated democracy'. I claim that aiming for the 'right values' in a centralised policy context inevitably leads to an 'expert' and top-down approach to value change. I will then highlight some aspects of an 'alternative report' which I will develop in my arguments for the prime importance of policies enabling the development of 'communities of environmental care'.

'Community' is, of course, a much contested concept and has recently become notorious as a 'feel good' prefix for potentially unpopular policies. The sense in which I use it here derives from the educational concept of a 'community of inquiry', defined as 'any group . . . engaged in a philosophical exploration of ideas raised and shared with the group' (Fisher 1994a), but includes, importantly, the element of shared geographical location or 'locality'. I am not assuming either that local communities have a continuous historical existence or that they necessarily exist because of the geographical proximity of individuals. I am claiming that individuals who come together to debate local environmental issues form a 'community of inquiry'. This community may be limited to those with similar views, or it may be more wide-ranging to include many different viewpoints.

The concluding sections propose that substantive knowledge of moral traditions can best be viewed as providing valuable resources for social debate in reflective communities, and present the case for forms of processual moral expertise which can contribute to the enrichment of civil society. I argue that the development of reflective communities can help justify faith in the ability of a revived local democracy to generate environmental values worthy of respect. I claim that this approach can provide a solution to the democratic dilemma outlined earlier in this chapter and refer to the Local Agenda 21 process as embodying the potential of this approach.

STATEMENT OF THE 'CLASSIC DILEMMA'

The classic democratic dilemma is between the 'right values' and the 'right processes' in public policy – of course we want both, but in pursuing either one we may lose the other. Often, with Plato, we do not trust democratic processes (the 'right processes') to yield values which we can respect, and if we pursue the 'right values' we may sacrifice the majority agreement necessary to enact these values in democratic societies. For the 'democrat' the 'right processes' have to be processes which confer political legitimacy through the involvement of citizens. Within democratic theory and practice there is a broad range of degrees of involvement from representative democracy (least) to direct democracy (most). There is, of course, much debate about which level of involvement constitutes *real* democracy.

To turn to the second horn of the dilemma, for the moral philosopher the criteria for deciding on the 'right values' are equally contested. However, there is almost universal agreement to the proposition that 'something is not right just because the majority believe it so' (Plato's *The Republic*; Snare 1992). In moral philosophy there is always, at the very least, a need for good supporting reasons why something should be considered to be 'right' – majority support does not and *should not* weigh in the moral scales.

THE CURRENT RESOLUTION OF THE DILEMMA IN THE UK: 'MITIGATED DEMOCRACY'

An example of the classic dilemma in the UK can be found in discussions of capital punishment. It is often claimed from opinion poll evidence that if there were to be a direct vote (referendum) of all citizens on the introduction of capital punishment for some offences there would be a majority in favour. It is then argued that this would be a morally wrong policy; for example, for reasons of irreversible gross injustice in the case of wrongful conviction. Some opponents of capital punishment then add the argument that it is part of the job of parliament to prevent irreversible gross injustice – not simply to enact the most popular policies. In this way a general thesis emerges that democracy must be *mitigated* by established considerations of justice, ie the 'right values'.

In the above case the mitigating 'right values' are drawn from the legal system which has developed historically on the basis of concern for individual and corporate human rights to life and property. As can be observed, this conceptual system does not readily accommodate environmental concerns. I do not propose to enter into arguments about the value of the human rights supposedly enshrined in case law. My point here is that there is a paucity of general established legal principles respecting environmental values that can be called on to mitigate democracy in the UK at present.

The report on proposed environmental reforms (which I will be considering shortly) is doubtless designed to remedy this shortcoming by providing such principles. It is perhaps a question for the sociology of law to consider whether basic legal principles do ever, in practice, actually become established as a result of such reports. Certainly in the USA, for example, academics can have a direct influence on legal philosophy – but I do not believe this to be the case in the UK. Irrespective of the practical likelihood of this mode of influence, the question I wish to ask is whether this is a desirable model of environmental value change. I will argue that, for environmental values at least, there is another model which can solve the democratic dilemma in a more fruitful fashion and which further has a chance of practical application.

A further form of ethical 'mitigation' has recently been gaining ground in the shape of 'ethics committees'. Committees including 'experts in ethics' are increasingly called on to provide policy evaluation. It should be noted that expert committees are generally expected to work within the framework of existing law and thus cannot initiate overall changes in legal philosophy. Expert committees usually include representatives of religious groups who can be said to represent the moral views of these groups in so far as they are all orthodox. I have argued elsewhere (Parker 1994a) that moral philosophers cannot have the same relation of representation to the lay public, and most importantly cannot claim 'expertise' in delivering the 'right values'. Hence, in a secular society we simply do not have access to any privileged set of the 'right values'. A consequence of this, it seems to me, is that we must develop an account of changing social values which includes a more clearly articulated role for the moral philosopher, consistent with the philosophical values of reflection and debate.

ENVIRONMENTAL VALUES AND THE DEMOCRATIC DILEMMA

Democratic values are often used as foundational in discussion of values in policy, but environmental moral reformers want to change these values themselves. Typically they wish to incorporate concern for future humans, other present and future sentient organisms and/or ecosystems. There is much debate on the best and 'correct' formulation of these concerns, but as I do not intend to go into details here, they are referred to collectively in this text as 'environmental values'. Used in this wide sense 'environmental values' importantly include vital debates on sustainability and values (Engel and Engel 1990) highlighted in the Rio Earth Summit 1992 and increasingly high on domestic political agendas worldwide.

Proposals for the incorporation of environmental values into public policy take the following forms:

1. Absolute prohibition of some consequences of action, eg extinction of any species.
2. A requirement (eg in planning law) to weigh environmental values in the assessment of schemes.
3. The requirement to introduce advocacy for environmental values into democratic decision making.

Both 1 and 2 would, if enacted, reduce the autonomous sphere of activities of citizens and hence require political justification. The introduction of advocacy, 3, seems a democratically uncontroversial measure as it relies on persuasion within the established democratic system. However, this may be of some concern to democratic environmentalists if the advocacy is only envisaged as taking place at parliamentary and select committee stage. Hence, the key point in relation to 3 is the extent of the requirement for advocacy.

I argue that a crucial part of the task of environmental moral reformers should be to establish processes that can be widely accepted as both politically and ethically valid to change the current consensus on democratic values. In my view, this must include a recognition of the ways in which the current consensus embodies the results of power negotiations (Benhabib 1992). Those moral reformers who wish to argue for environmental values continuous with human welfare should state explicitly how autonomy figures in their account. I will present arguments to show that environmental values will be better served by extended democracy rather than by a managerial paternalism which minimises the positive role of social involvement in policy.

THE 'EXPERT' MODEL OF VALUE CHANGE

I now turn to the model of value change exemplified in the main report contained in the *Values, Conflict and the Environment* report (Attfield and Dell 1989) (References in this section will be to this document unless otherwise stated.) I focus less on the particular policy recommendations made and more on the model of value change employed in this report.

The report aims to produce at least a 'rough estimate of the right values put into an argument of the right form' (p41). The authors review 'a range of arguments which appear recurrently in environmental disputes' (p3) and these are drawn from the range of experience of the interdisciplinary group plus particular reference to two public inquiries. A range of environmental values from primary academic textual sources is also reviewed (p30). The proposed method, 'comprehensive weighing', is an extended version of cost–benefit analysis which takes into account human 'informed preferences' (obtained by consultation and/or estimation, p36) and the interests of all sentient life when deciding on environmental policy matters.

This report is an extremely valuable document but, I claim, exemplifies the shortcomings of the 'expert' model of value change. This model supports a notion of 'moral expertise' which relies on primarily textual information regarding the moral traditions of society. The process of enquiry then becomes one of submitting social problems to these traditions for guidance. The debate is therefore restricted to questions of definition of similar cases in order that traditional theory can be extended to cover these cases (see Chapter 4 for a more detailed account). There is an assumption that new situations are fundamentally continuous with past situations. It is implied that in making a synthesis of these moral traditions some kind of publicly acceptable consensus will be produced. Dissemination of results is viewed as a primarily top-down or 'cascade' model, hence the main problem is to convince decision makers at a high level in an established system – it is only later that public education is required.

The proposed public involvement in 'comprehensive weighing' is restricted to the feeding of 'informed preferences' into the weighing process. Clearly someone has to decide to classify some responses as 'informed' and others as 'uninformed' and therefore suitable to be ignored. There seems to be no reason internal to the theory why this process could not be carried out by civil servants without any other kind of public participation. Most importantly, no detailed consideration is given to the degree or kind of public involvement necessary to validate comprehensive weighing itself. While the report agrees that a referendum of the 'affected population' might be held on site-specific issues, it is stated that this is 'not practicable for matters of policy' (p59). Equally, there are calls for community moral education in 'observing corresponding informal rules', but public involvement is seen only as *instrumental* in ensuring that the theory is applied. The same instrumental perspective is found in the support given to local government in the report.

There is no doubt that the editors and contributors to this report are genuinely concerned to include public debate as part of political processes in general. They advocate a Standing Commission to report to all sections of the population on available options for energy generation, for example. They believe that the findings of such a commission should not be presented as a 'best buy' but rather in a dialectical fashion with arguments for and against any course of action. It is ironic that their view of policy is based on a privileging of expert opinion such that comprehensive weighing is itself presented as a 'best buy'. This reflects the tensions within the report as to its own dialogic nature. On the one hand it is presented as a resource to be consulted 'by all those who find themselves involved in conflicts of value concerning environmental issues' (p1), but on the other hand the report implies that questions of value conflict are too complex to be openly considered in public fora. The other aspect of this tension is found in a section regretting the progressive restriction of local government in the UK. The argument here does not directly challenge the policy hegemony of centralised government, but refers vaguely to 'omissions' of viewpoint from 'appropriate' public bodies. In fact this main report cannot consistently challenge centralised policy, as it is precisely this mechanism through which comprehensive weighing is envisaged as working.

This report and any approach which privileges a 'right values' perspective can only logically result in calls for public moral 'training'. If the 'right values' are decided by 'experts', then all that remains is to train the populace. This is in direct contradiction to the widely recognised need for communities to 'own' their values. In this model there is no recognition of public participation in environmental issues and the accompanying debates as independent and valuable sources of developing new values.

THE TECHNOCRATIC MODEL

The expert practice described above comes to have an unfortunate resemblance to the technocratic model of social value change (Lowe, Clark and Cox 1993). In the technocratic model, social objectives are decided by economists and social scientists against the background of an unquestioned social power structure. 'Moral philosophers' are then employed to sell the corresponding 'values' to the public. Philosophers who find themselves in this position are being employed as propagandists, not as reflective practitioners of applied philosophy. The similarities with the 'expert' model come in the methods of public persuasion and training; once the 'right' values have been decided, whether by philosophers or by technocrats, the resulting attitude to the public is the same.

THEORY AND PRACTICE

The value of this report is found partly in the articulation and consideration of values expressed by those involved in actual environmental disputes. Thus, in this way, some part of the achievement of non-governmental organisations (NGOs) in generating moral debate is recognised in practice, if not explicitly stated. I argue that the failure to acknowledge overtly the continuing role of social movements in challenging values is a natural consequence of failing to challenge an elitist view of theory. The expert approach itself relies on a static metamodel of the relation of theory to practice (Figure 3.1a). This model accurately depicts the top-down assumptions of most policy review processes. Currently, reviews take place within a time constraint. Resulting guidelines then endure until enough pressure builds up among practitioners and concerned organisations for a new review. Overall this results in the wasting of human energies and generates conflict. It points to the necessity of creating the responsive and varied culture of policy review suggested below in Figure 3.2.

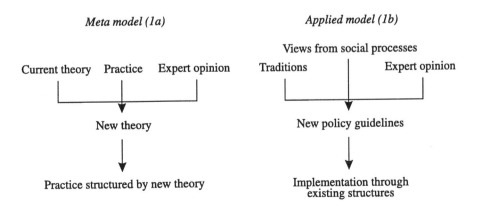

Figure 3.1 Expert model

THE 'ALTERNATIVE ACCOUNT'

The alternative account by Robin Grove-Oliver and Oliver O'Donovan is printed as a 'minority' report within the *Values, Conflict and the Environment* document. This alternative account views the important site of moral consensus as being within the community itself. The authors propose that those who care about the local environment should receive a better hearing as they have 'distinctive understandings and insights to share' (p80). These insights are 'essential elements in the deliberations of a community', helping the community 'identify and articulate its values' (p81). This account does not employ textually embodied traditions as validating any particular set of values, but it does argue for prioritising 'irrecoverable environmental goods' (p80).

One of the weak points of the alternative account is the vague description of the desired social processes. If those who 'care' for their environment are to be given preferential hearing, how will they be picked out from other local citizens and by whom? It is not really even clear from the alternative approach if the authors mean to empower preferentially local decision making; if so, the recommendations they do make for prioritising environmental goods stand in contradiction to this strategy. In summary, the alternative account wants it both ways. It is simply inconsistent to uphold the role of social processes in generating values and still to maintain the 'expert' prerogative recommending legislation without public debate.

Perhaps part of the problem here is that it is, of course, impossible to *justify* the as yet unknown values to be yielded by these enhanced processes. In this account, it is the faith in the social processes themselves which requires justification, and I do not find any attempt to provide this in the text. Finally, this view does not provide any kind of positive role for theory or for reflective practitioners. I believe both that there is much of value to be gleaned from theory and that philosophical skills can play a catalytic role in enabling moral reflection. It is to this that I now turn.

MORAL EXPERTISE

I have stated previously in this chapter that there can be no moral expertise in the sense of delivering the 'right values'. I here elaborate this statement and argue for two kinds of limited but legitimate forms of moral 'expertise' – substantive and processual. My arguments against moral expertise have been twofold. First, I have argued that the notion of moral expertise stands in contradiction to the important principle of moral autonomy. Secondly, I have argued that moral philosophers cannot be said to *represent* the moral views of the lay public. In respect of that second argument I must now add the point that moral philosophers cannot claim to represent uncritically the views of the public and uphold the principle that majority views are not held to be more valid than minority views. Sociologists studying values might also wish to add that the representative's role would require knowledge of the actual distribution of people's values, which most philosophers make little attempt to discover in any systematic way.

The two arguments above against moral expertise must be complemented by a third based on the relation of the 'expert' to a body of knowledge. I have argued at greater length elsewhere (Parker 1994a) that our notion of expertise in professions includes reference to an agreed body of knowledge which the experts employ in their practice. It might be claimed that moral philosophers have special knowledge of a body of texts embodying moral traditions and the fruits of previous moral enquiry. However, the relation of moral philosophy to these texts is one of continual criticism, reworking and re-evaluation. Questions around what constitutes the 'moral tradition' itself are continually raised, by feminist ethicists to cite one example. Healthy and active philosophy presents a diverse and continually changing field of enquiry, not a body of expert knowledge. In spite of this people may, and often do, protest that the study of moral philosophy must surely enable philosophers to contribute to the moral understanding of the wider public in *some* way. I consider below the two ways in which I believe moral philosophy can legitimately offer to enable and enrich public moral understanding.

SUBSTANTIVE MORAL UNDERSTANDING

As a philosopher I do want to argue, not surprisingly, that familiarity with a body of texts dealing with moral questions does represent useful knowledge of a particular kind. This study provides an awareness of the variety of approaches which have been used by those reflecting on moral questions in their writing, awareness of the historical context of much current thinking, and an understanding of the implications and associated commitments of various ethical positions. Many moral philosophers are fruitfully employed in exploring theoretical questions which may only be of direct interest to those who are prepared to pursue the subject in depth. In their case it would not be appropriate to demand that this knowledge should be publicly accessible. However, I argue that moral philosophers who enter into the realm of public policy need to guard against public misapprehension of their substantive moral knowledge as 'expertise' by making their knowledge publicly available.

I propose that the remedy for the problem of 'expertise' can be the adoption of an ideal of 'public service' in attempting to communicate knowledge of moral traditions to provide a *resource* for public deliberation. This social role involves a clear separation between one's own moral views and commitments and the responsibility to provide the public with a range of ethical resources to enable public moral reflection. There is an understanding of these two roles in teaching moral philosophy to students – no course would be considered up to standard which was based totally on the lecturer's own moral views.

Philosophers sometimes appear to be less scrupulous in making this distinction in roles when making public interventions.

In my analysis of *Values, Conflict and the Environment* I have claimed that moral knowledge is used in a primarily top-down attempt to change values. I argue that the ideal of public service in providing accessible ethical resources is supportive of democratic value of moral autonomy and is both an ethically preferable and a more realistic method of engendering value change. This recommendation is more realistic because it does correspond with the ways in which moral arguments are used in the public context. For example, Kate Burningham (see Chapter 7) gives an account of the use of ethical resources in public discourse about road-building schemes. In the study cited the resources used by participants in the debate to argue ethical points were in many ways inadequate to defend local action in protection of the environment. This fact serves to underline the need to enable greater public access to a wider variety of ethical resources if we genuinely wish to enable environmental value change.

'Access' is the key word here: in order to render ethical resources for discourse publicly accessible effort has to be made to present them clearly and simply. As a community educator, I have learnt that many people unfamiliar with philosophy are very capable of using complex concepts but not complex language. However, knowledge of resources is not, in itself, sufficient. Skills must be acquired in using resources appropriately and effectively, and it is to this more processual aspect of moral practice to which I now turn.

PROCESSUAL EXPERTISE

Separating method from content is always problematic, as substance and process are always related. I would argue, in this context, that there is a sufficient distinction to be made between substantive areas of moral philosophy as outlined above and moral philosophy as practised, for example, in critical thinking (Fisher 1988; Paul 1989). The critical thinking approach concentrates on developing the skills of the individual student and his or her capacity for autonomous moral reflection. Moral content enters in the ground rules of respect for others' opinions – in fact, the necessary minimum to allow debate to take place. Processual expertise in ethics covers skills such as analysis and presentation, methods of argument and assessment, and knowing how to use the kinds of conceptual structures which can aid moral reflection. This is 'expertise' in the same sense that one can become an 'expert' at using certain tools.

Skills develop through being practised and experience of moral debate is always a necessary part of developing moral autonomy. This is recognised in the UK Further Education Unit document on moral education, for example, which contains suggestions for activities to develop moral autonomy in addition to a strong substantive moral content (FEU 1989). I now propose to consider in more depth the social practice of moral autonomy and its relation to decision-making processes.

MORAL AUTONOMY IN REFLECTIVE COMMUNITIES

In the West we tend to see moral reflection as an individual activity. We particularly admire the moral reformer in heroic mode struggling against the social consensus; we do not really have a well developed complementary image of social moral reflection. I have already referred to what is, in my view, our lamentable tendency to allow social moral reflection to be dominated by small groups of 'experts'. I now wish to argue for the alternative practice of moral reflection in communities.

I have already suggested above that effective moral reflection in communities requires ethical resources to be available and that the quality of that reflection will be affected, partly at least, by the degree of processual moral expertise available. A key objection to this strategy may be made on the basis of moral autonomy. It may be claimed that in order for communities to be truly 'moral' they must be composed of individuals who are morally autonomous (Fox 1986). If this point is accepted the faith in social processes can be undermined in the following way: it is claimed that we know that most communities are not made up of such individuals and therefore we cannot trust values produced in such communities.

I argue that the existence of a reflective community is a necessary precondition for individual moral autonomy and not the other way around. In order to develop my argument I need to look more closely at different aspects of the concept of 'autonomy'. One aspect which is particularly stressed in medical ethics is autonomy as 'self-determination' (Gillon 1985). In medical ethics the concern is that medical practice should not trespass on the self-determination of patients. However, in ethics itself we are concerned with what it means for a person to be *ethically* self-determining. Michael Fox has argued (Fox 1986) that moral self-determination centrally includes the capacity to use complex concepts and to accept responsibility for actions. I want to argue that both of these capacities are generated by involvement in morally reflective communities.

I have referred above to the need for ethical discourse to be readily available in social fora. I further suggest that individuals need access to these concepts in a context in which their social meaning can be understood. It is the lack of this context of practice which diminishes the value of some academic courses in moral philosophy. There is much evidence, for example from the practice of philosophy with children (Fisher 1994b), that human beings in groups do in fact spontaneously generate moral concepts. However, this process is facilitated and encouraged by the provision of a specific social space in which this can take place and by the presence of a facilitator (Daly 1994). My main point here is that the ability to use complex moral concepts is developed by engagement in social moral reflection.

The ability to take responsibility for actions is the other proposed key feature of moral autonomy. Taking responsibility for actions is an 'ability' which in one sense cannot be tested except in practice. We are never really sure if anyone will 'rise' to the reponsibility with which they have been entrusted. The way we usually develop this ability is gradually to increase responsibility – sometimes we learn responsibility for our actions the hard way. In any event there is no substitute for the experience of actually having responsibility, and there is plenty of evidence that when people know that their opinions are decisive, in the Swiss referendum system for example, they do rise to the occasion by engaging in detailed and responsible debate.

In summary, I claim that the existence of morally reflective communities is a prior condition for moral autonomy and hence for attempts for moral reform through argumentation. These reflective communities form whenever people come together to discuss issues, but are facilitated by the provision of social spaces, fora and structures within which this reflection can take place and by the presence of facilitators ensuring the minimum requirements for debate. I conclude that an important and currently underdeveloped form of moral expertise could be the development of structures and skills that enable the moral reflection of communities, understood as local civil society.

COMMUNITIES OF ENVIRONMENTAL CARE

I now wish to turn to the proposal in the alternative report that those who care for their local environment have 'distinctive insights to share'. In order to support my call for greatly devolved environmental decision making I will make out a stronger case for the ethical knowledge of communities in respect of their local environments. I use 'ethical knowledge' in this context to mean 'awareness of value' and my argument is based on the specificity of local environments. I make the overall value judgement that our public policies need urgently to improve environmental care.

In Chapter 1, Michael Redclift has pointed to the difficulties in achieving international consensus on global environmental care and the variety of value-related considerations that determine the interpretation of scientific information. In the light of these difficulties some philosophers have attempted to facilitate moral consensus by creating and arguing for some new overarching theory of environmental values on which to base policies for environmental care. However, the social and intellectual development of environmental concern has been driven by a great variety of different values found in the diversity of life (Brennan 1992). At least one important feature in this process has been the concern of local people for the distinctive environments which they inhabit (Clifford and King 1993). This concern is not only manifest in rural and privileged urban environments. People in deprived urban environments often have a strong sense of what is needed to improve their environments and local participation in designing the local environment increases local identification (Turner 1990). Policies enabling responsible local concern should be built on the recognition that the value of localities to their inhabitants can form a powerful motive for local environmental stewardship.

Opponents of this view may fear that there will be a loss of common interest and concern if communities are empowered in environmental decision making. At present the supposed 'general benefits' of development schemes can be much more clearly stated than can the general loss of specific environments. It is at this point that the local awareness of the value of the specific environment under threat is quite literally irreplaceable. Who else, except those who actually live there, can know in depth the value that is to be found in living life in that particular place? Some would claim that the loss of a specific environment is merely a loss for those who live there and should therefore count for less in the ethical scales than should the supposed general gain. To argue this way is to assume that to care for something specific is always less 'ethical' in some way than to care for something more general.

I have argued elsewhere (Parker 1994b) that to care *particularly* for something is an ethical response that can, itself, be a generalised moral principle. That is, we can agree that caring particularly for some things is a universal part of ethical life. Mary Midgley (Midgley 1993) has described this as the sphere containing those valuable things to which we as individuals are deeply attached. I propose that building environmental stewardship around those deeply felt and ethically legitimate local concerns is a viable strategy. This process will require a continuing dialogue between local value narratives and decisions and global policies and priorities. I explore the role of Local Agenda 21 in developing this dialogue in my concluding section.

THE DEMOCRATIC DILEMMA AND COMMUNITIES OF ENVIRONMENTAL CARE

I now return to my original enquiry into the democratic dilemma to claim that the proposal to enable the development of morally reflective communities can help solve this dilemma. As we have seen above, an essential ingredient of the dilemma is the belief that we cannot trust democratic processes to yield values which we can respect. I want to argue that we *can* trust *reflective* democratic processes to yield values which we can respect. It is important to aim for values which we can *respect* rather than values which we can *agree* with. Disagreement and development are an integral part of the reflective moral life and therefore in social matters we must aim only for values which we can respect.

Many people view the prospect of extended democracy with alarm as they envisage this within current structures. For example, people tend to imagine an unreflective public voting in a referendum on issues which they do not understand. The proposal here is the extension of local democracy to include greater autonomy in environmental decision making. I have claimed that in matters of local environmental care it is the local community which is best placed to judge, both ethically and practically. I claimed also that responsibility has to be developed through practice and experience. I further argued that the institution of reflective, locally based processes should be a condition of this greater autonomy. I am not claiming that local processes will automatically be perfectly democratic or completely ethically responsible. I am claiming, however, that they would be both more democratic and more ethically responsible than the current processes defined above as 'mitigated democracy'. My proposals are based on a process metamodel of the relation of theory to practice, as shown in Figure 3.2a. Figure 3.2b shows the interpretation and application of this as expounded in this chapter.

This application importantly includes the notion of reflexive assessment of new theories and endorses a plurality of practices and approaches. I maintain that a proliferation of local practices will be invaluable provided they are evaluated and assessed by practitioners and theoreticians/policy makers. The circumstances in which we find ourselves – the need rapidly to develop more sustainable ways of living, incuding wide social changes – are such that a degree of creative experiment is required. My premise is that, if we wish to maintain democracy, this cannot be an experiment conducted from the top by technocrats from any profession.

LOCAL AGENDA 21

My final section argues that the kind of local process outlined above is necessary to enable the development of stewardship towards global sustainability. Local Agenda 21 has been envisaged as a process which engages all sections of a local community (loosely defined as 'local' to include, at the time of writing, both borough and county councils in the UK, for example). The remit of Local Agenda 21 is extremely wide but should involve the community in consultation and debate to develop Local Agenda 21 commitments, mirroring and extending

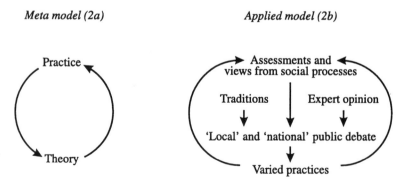

Figure 3.2   Process model

those international agreements obtained at the Rio Earth Summit in 1992 (Quarrie 1992; Keating 1993). This includes the development of local plans for sustainability as a counterpart to those plans being prepared at national levels. The completion date for this extremely ambitious exercise is set for the end of 1996, but clearly this can only be the first stage of a learning and development process which must extend into the foreseeable future. I want to consider here the dialogic opportunities offered by this process for environmental value articulation and change.

Local Agenda 21 has been taken up in a variety of ways across the UK. Some councils are simply dusting off existing 'green' policies and re-entitling them 'Local Agenda 21'. At the other extreme, some areas have embarked on a genuinely extensive programme of networking, debate and discovery. Local Agenda 21 has the potential to bring together for the first time local members of NGOs who are working in different areas of concern to discuss a common agenda. One tension and development produced by this local focus is to change the emphasis from centralised NGO decision making to one which stresses local contributions from members towards local solutions. Centralised organisations are thus challenged to provide an enabling rather than a directive role. Part of the concern of this chapter overall is to enquire into the possibility and implications of adopting an enabling role to help environmental value development.

Local Agenda 21, where it is being developed as it should, provides all the necessary conditions for the development of communities of enquiry focusing on the key questions of the relationship between human and environmental welfare. The potential for developing reflective processes which could help clarify issues and aid local policy decisions is enormous, particularly in view of the continuing nature of these local debates on sustainability. The creation of communities of enquiry about local sustainability can contribute to the regeneration of local civil society which so many environmentalists regard as essential for environmental protection (Parker 1994c).

In summary, I argue that this local process should be prioritised over the development of centralised policies for the following reasons:

1. Western society is desperately in need of mechanisms to enable 'primary environmental care'. This cannot be achieved by the highly expensive and ineffective method of centralised legislation policed by paid inspectors. The alternative is to develop the capacity for effective control at the local level.

2. Centralised policies are inadequate to respond to complex local conditions, create resentment and alienation and currently favour general 'development' over local 'conservation'.

3. Engaging in policy development debates is the best kind of active education for sustainability which effectively utilises local knowledge.

4. To enable real change we need to generate a diversity of alternatives from which to learn new ways of living sustainably.

5. Enabling local moral autonomy is ethically preferable to imposing values through centralised policy and ensures the 'ownership' necessary for values to be enthusiastically championed by communities.

To conclude on a philosophical note, I woud like to quote from R. M. Hare's characteristically understated summary of his views on moral reasoning in a recent interview (Johnson 1994), 'It is a question of whether we are going to be Platonists and say "only a few of us can do this properly and the rest have to do what they say" or whether we are going to be more liberal and say, "well, most of us can in fact do it after a fashion, and actually the high-ups aren't much better at doing it than we are, so we had better all do it."'. To this I can only add, 'where sustainability is concerned we'd better all start doing it where we live'.

REFERENCES

Alcoff, L. and Potter, E. (eds) (1993) *Feminist Epistemologists*, Routledge, London.

Attfield, R. and Dell, K. (eds) (1989) *Values, Conflict and the Environment*, Ian Ramsey Centre, St Cross College, Oxford.

Benhabib, S. (1992) *Situating the Self*, Polity Press, Cambridge.

Brennan, A. (1992) 'Moral Pluralism and the Environment', *Environmental Values* **1**, 15–33.

Clifford, S. and King, A. (1993) *Local Distinctiveness*, Common Ground, London.

Daly, J. (1994) 'Philosophical Enquiry', paper presented at the Critical Thinking Conference, University of East Anglia.

Engel, R. and Engel, J. G. (1990) *Ethics of Environment and Development*, Belhaven, London.

Fisher, A. (ed) (1988) *Critical Thinking,* Proceedings of the Conference held at University of East Anglia.

Fisher, R. (1994a) 'Stories for Thinking', paper presented at the Critical Thinking Conference, University of East Anglia.

Fisher, R. (1994b) *Centre for Thinking Skills Information Pack Vol 3*, Centre for Thinking Skills, London.

Fox, M. (1986) *The Case for Animal Experimentation*, University of California Press, Berkeley.

Further Education Unit (1989) *Moral Competence*, FEU, London.

Garry, A. and Pearsall, M. (eds) (1989) *Women, Knowledge and Reality*, Routledge, London.

Gillon, R. (1985) *Philosophical Medical Ethics*, Wiley, Chichester.

Johnson, A. (1994) 'Applied Philosophy and Moral Theory', *Philosophy Today* **17**, September.

Keating, M. (1993) *Agenda for Change*, Centre for our Common Future, Geneva.

Lowe, P., Clark, J. and Cox, G. (1993) 'Reasonable Creatures', *Journal of Environmental Planning and Management*, **36**, 1, 101–111.

Midgley, M. (1993) 'The end of Anthropocentrism', paper presented at Royal Institute of Philosophy Conference, Cardiff.

Mill J. S. (1974) *On Liberty*, Penguin, London (first published 1859).

Parker, J. (1994a) 'The Professional Ethics of Moral Philosophers on Expert Committees', in R. Chadwick (ed) *Professional Ethics*, Avebury, Hants.

Parker, J. (1994b) 'Disabling Ethical Resources', working paper.
Parker, J. (1994c) 'How Must our Values Change for the Sake of Future Generations', discussion document, WWF, Surrey.
Paul, R. (ed) (1989) *Critical Thinking Handbook,* Sonoma State University, California.
Quarrie, J. (ed) (1992) *Earth Summit* '92, Regency Press, London.
Snare, F. (1992) *The Nature of Moral Thinking*, Routledge, London.
Turner, B. (ed) (1990) *Building Community*, for Habitat International Coalition, from the Hastings Trust, 12 Wellington Square, Hastings, E Sussex.

4 A Green Thought in a Green Shade: A Critique of the Rationalisation of Environmental Values[1]

MICK SMITH
Department of Applied Social Sciences, University of Stirling

Today, the 'valuation of the environment' has become a purely technical problem, an arena for the machinations of environmental economists or moral philosophers. Yet the all too ready acceptance of this role by institutionally appointed 'experts' exhibits a lack of reflexiveness on their part. They seem unaware of the limitations that their intimate relation to, and inclusion within, dominant bureaucratic social structures imposes on their theoretical pronouncements and methodologies.

Reflexivity demands that we recognise that our theoretical expressions are inscribed within the practices of our own social formation, within the past traditions, present circumstances and future hopes of *this* society. The academic sphere, mimicking the structure of society at large, imposes its own constraints on both the form and the content of the theoretical constructs in which it deals. For example, papers are first delivered, verbally or in print, as monologues and then defended in an adversarial atmosphere against all criticism. This is typical of a climate where career-oriented individuals (of which the environmental sector has more than its fair share) compete against each other for intellectual capital. This format frequently leads to our talking past rather than talking to each other. I make this point simply to illustrate how any theoretical communication will inevitably be influenced by sociological and political factors, and to remind us that even interdisciplinary conferences are, to a degree, microcosms of a hierarchical and managerial society moulded by the same temporal and structural forces which are destroying the environment.

In his 'Theses on the Philosophy of History' Walter Benjamin describes a:

> painting named *Angelus Novalis* [which] shows an angel looking as though he is about to move away from something he is fixedly contemplating. His eyes are staring, his mouth is open, his wings are spread. This is how one pictures the angel of history. His face is turned towards the past. Where we perceive a chain of events, he sees one single catastrophe which keeps piling wreckage upon wreckage and hurls it in front of his feet. The angel would like to stay, awaken the dead, and make whole what has been

Values and the Environment: A Social Science Perspective, Edited by Yvonne Guerrier, Nicholas Alexander, Jonathan Chase and Martin O'Brien. © 1995 John Wiley & Sons Ltd.

smashed. But a storm is blowing from Paradise; it has got caught in his wings with such violence that the angel can no longer close them. This storm irresistibly propels him into the future to which his back is turned, while the pile of debris before him grows skyward. This storm is what we call progress. (Benjamin 1992, p249)

Radical environmentalism is a moral and political protest against this seemingly irresistible hurricane of destruction, a protest which recognises that the problems of deforestation, ozone depletion, loss of habitats etc cannot be treated in isolation. They stem from and are entwined with our modern forms of life. The critique of environmental destruction necessarily becomes a critique of contemporary society.

How are we to carry forward this critique? If the problem lies deep in our society then we must recognise that the theoretical language we have available may itself prove to be tainted by a form of life which depends for its very existence on the continuing environmental holocaust. As Alasdair MacIntyre points out, there are serious implications for those ethical theories which attempt to define solutions to our current crisis *within* philosophical frameworks which arise from and whether knowingly or unconsciously support the society we wish to criticise.

The ability to respond adequately to this kind of cultural need depends of course on whether those summoned possess intellectual and moral resources that transcend the immediate crisis, which enable them to say to the culture what the culture cannot say to itself. For if the crisis is so pervasive that it has invaded every aspect of our intellectual and moral lives, then what we take to be resources for the treatment of our condition may turn out themselves to be infected areas. (MacIntyre 1981, p3)

In environmental ethics – the field with which I am most concerned – one can recognise a range of responses to our crisis. Some, it must be admitted, do not even recognise the relevance of MacIntyre's point. They are content to operate within the safe confines of the accepted paradigms of moral theory, whether utilitarian or rights based, and continue to view the non-human world as of only instrumental value. Their anthropocentric and blinkered perspective refuses even to recognise the empirically obvious – that many cultures other than our own, and many people within our own society, do indeed have genuine *moral* concerns for our non-human environment. We see the destruction of whales and forests as an evil comparable with crimes against humanity. To treat sacred groves, rare insects or even the tree we used to climb as a child as potential resources is both an appallingly reductive misunderstanding of their complex relations to ourselves and morally on a par with offering to sell one's grandmother to the highest bidder.

By ruling moral consideration for the environment itself out of court this philosophical parochialism colludes with those who seek a purely economic solution to environmental problems. Value-rational, traditional and affective actions are all subsumed under a system of instrumental rationality, in this case an economic system which focuses on the relative efficiency of particular means while claiming to be neutral with regard to the determination of ends. (This terminology is that of Max Weber who refers to the process of rationalisation – the spread of instrumental techniques of economisation and bureaucratisation – as bringing about the disenchantment of the world and the 'iron cage' of the totally administered society.) In fact this reductive economisation is certainly not neutral but a powerful influence which imposes its own *raison d'être*, its own instrumentality on all spheres of life. The invisible calculus of dollars and yen in tandem with a visionless bureaucracy steamrollers all opposition based on an attentive understanding of our complex relations to each other and the natural world. Anything and everything is simply assumed to be a potential or actual resource for human use.

Not all ethicists are quite so dismissive of the environment. Some, perhaps the majority of those in academic circles dealing with environmental ethics, are a little more flexible. They take on board MacIntyre's point in so far as the *content* of their ethical deliberations is concerned, but refuse to recognise that the *form* of these theoretical structures might also be implicated. These people (who include figures such as Peter Singer, Tom Regan, Robin Attfield) are genuinely concerned to expand the boundaries of moral considerability beyond the human horizon, yet their methodology remains almost entirely unaltered. They simply look to the 'natural' world for novel grounds on which we might found a theory of animal *rights*, or argue for the inclusion of a selected few nonhumans into the machinations of a utilitarian calculus. This position, which might be referred to as axiological extensionism, has almost limitless potential for keeping moral philosophers employed. For example, exam questions can now be set asking 'Are bacterial rights infringed whilst breathing?' or 'How should one recognise the best interests of rivers?' and so on. However, in practical terms this methodology is about as beneficial for the environment as the current government's job clubs are for the British economy. Where instrumentalism reduces all values to values for humans, axiological extensionism becomes progressively more impractical the further one moves away from the human sphere (Smith 1991). Not only this but, as I shall argue, its very form still operates to reflect and reinforce our current social structures.

This chapter constitutes a challenge to the hegemony of these formal conceptual systems and methodologies and their claims to evaluate the 'natural' environment. Superficial debates about the moral considerability of non-human entities are academic in both senses of the word. Apparent differences of opinion, eg about which animals have rights or interests, mask a deeper and more fundamental conformity within contemporary philosophy. The environment simply provides a novel area in which to *apply* tried and tested ethical formulae. The content of environmental ethics might be different from that of medical ethics (deforestation rather than organ transplants), legal philosophy (extinction rather than execution) or business ethics (greenery rather than greed), but the form remains identical.

The dominant forms of moral theory attempt to provide a rubric which can be used to determine right and wrong by those not intimately associated with the circumstances – ie bureaucrats, governments, law courts etc. Ethics is given a role as a theoretical tool for passing judgements or evaluating actions. Such formal rubrics facilitate managerial and technical efficiency, eg in evaluating the ethical 'cost' of a road development or comparing the rights and wrongs of a quarrying operation. They promulgate and support the myth of a *neutral rationality* in the hands of professed experts as an impartial arbiter in the service of society as a whole.

While supposedly neutral, this conception of ethics implies a particular understanding of the relation between theory and practice – a relation by which theory claims to encapsulate and represent the essential features of moral activities and then re-apply them. Morality is reduced to a series of abstract formulae which can supposedly be applied to circumstances irrespective of the context of the moral claims involved. However, looked at differently such formulae act as an ideological smokescreen, giving the (false) impression that our moral concerns for the environment have been addressed, weighed in the incorruptible balance of rational thought and found wanting.

These explicit and formalised systems have come to colonise and dominate the modern life-world. In adopting this form current ethical theory operates as yet another form of positive philosophy – that is, a philosophy which supports rather than subverts the current *status quo*, philosophy as an instrument of social management rather than as an expression of genuine moral concerns. In Zygmunt Bauman's words, 'contemporary ethics embark[s] . . . on an arduous campaign to smother the differences and above all to eliminate all 'wild' – autonomous, obstreperous and uncontrolled – sources of moral judgment' (Bauman 1993, p12). Formal ethics misconstrues and eviscerates our actual moral feelings in order to incorporate them, as pale shadows of their former selves, into a hierarchical society where decisions are taken for us by others. In this way axiological extensionism, no less than its more blatantly instrumental and anthropocentric counterparts, defuses environmentalism's radical critique of Western society by marginalising those who speak with a different voice. This conception of ethics carries with it a self-justifying rhetoric based on its own pretensions to embody the one and only rational approach to moral valuation. So long as formal rationality dominates the ethical realm to the exclusion of all else, then anyone challenging the 'rational' form of such philosophy is simply deemed irrational and hence inconsiderable.

The work of Pierre Bourdieu can help both to understand how moral theory has come to take on such a restrictive form and to point the way towards alternative conceptions of morality. Bourdieu favours a problematic which is reflexively aware of its own origins, claiming that one cannot simply conceive of theory as a process of uncovering the truth, an *opus operatum*, but must rather see that theoretical practice (including that of moral philosophers) is itself a particular form of life – a *modus operandi*. The Wittgensteinian critique of objectivism, which has heavily influenced Bourdieu, suggests that theoreticians are 'prisoners' of *their own* cultural *milieu*, that of Western society in general and more specifically of a particular academic and managerial tier within that society. It is impossible for theoreticians to escape their own presuppositions, to take a position from nowhere. The moral theorist's explanation is always a construction via the theoretician's practical activity. This explains why other members of theoreticians' own academic and wider culture find their explanations convincing.

Bourdieu advocates a form of reflexivity whereby any theoretical field should come to see itself as a historically and culturally specific practice in a dialectical relationship with its subject matter. Any account of moral theory thus needs to take into account the theoreticians' own practical interactions with the material of their study and the propositions on which that practice is based. This entails the reflexive recognition that theories produced to explain actions, values etc incorporate the social structures of the *observer's* society as well as those to be *observed*. Bourdieu thus develops a sophisticated form of cultural relativism whereby the ideological presuppositions of a society determine, at least in part, the interpretation and understanding which can be achieved by that society.

Bourdieu's epistemological critique of positivistic theoretical understandings thus focuses on their inadequate conception of their own relations to the 'objects' of their study. In keeping with his reflexive epistemology of social theory, Bourdieu rejects the usual explanations of the regularities produced and reproduced in social formations. Such regularities are frequently explained in terms of 'rules' – underlying structures recognised and expressed by the theoretician. But, on Bourdieu's reading, the rules that moral theorists such as Singer et al claim to excavate can be interpreted very differently. They result from the unconscious imposition of the mode of juridical social regulation and formal rationality which dominates our own modern society onto their understanding of theory itself. Both the production of formal ethical codes and their misrecognition as 'logical' representations of moral values are the result of unreflexive anthropological practices which take no account of their own social origins. Theorists naturally and unthinkingly take such *rules* as a model for the operation of all other aspects of our society because our society is itself rule governed. Academics produce formal theoretical structures for determining values because this fits with their ideologically produced expectations about what ethical theory 'must' be like.

We can no longer retain this positivistic and representationalist belief in the objective position of the theoretician. Instead we should understand theory as a particular form of practical activity relative to our own social framework. Theory does not *represent* objects in a realm of pre-given form or logic and in a pre-given relation to the world, but is an expression of the practical interaction of theoreticians with their surrounding environment. Theorising is a practical activity which incorporates and objectifies the social (and environmental) structures of which it is a part both in its form and in its content.

If ethical rules and moral axiologies are an expression of modern social structures at the level of theory, how then are moral feelings and conduct to be understood? We need theoretical perspectives which allow us to oppose the moral disempowerment which positive philosophy brings in its wake. Bourdieu develops an alternative conception of rules similar to that found in the later Wittgenstein, a looser conception whereby rules are no longer to be envisaged as consciously formulated limits on social action but rather as ideologically incorporated and open-ended *strategies*. Strategies, unlike explicit rules and formal systems, are not concrete artifacts applied to situations by rote. The members of any given society would usually have no need of such rules, instead they have a feel for the social field in which they exist, an 'unconscious' ability to act within the expected bounds of that field; just as, for example, a footballer may exhibit a mastery of the game by knowing when a pass or a shot at goal is likely to succeed without ever following explicit laws on such matters.

This 'feel for the game' is incorporated into individuals through their immersion in society. Our behaviour is not to be understood as driven by hard and fast laws but as the product of dispositions inculcated in the earliest years of life and constantly reinforced by calls to order from the group, that is to say from the aggregate of the individuals endowed with the same dispositions, to whom each is linked by dispositions and interests (Bourdieu 1991, p15). The observer steeped in our social ethos may express the perceived patterns of these dispositions in terms of rules, but no such rules actually exist – one must not make the mistake of reifying the results of theoretical practice as existing entities. Where explicit rules do exist in other societies, they normally exist only as a second line of defence intended to make good the occasional misfiring of the collective enterprise of inculcation (Bourdieu 1991, p17).

This system of dispositions transmitted from generation to generation is referred to as the 'habitus'. This generative habitus is, Bourdieu explains (p9), a series of dispositions acquired through experience, thus variable from place to place and time to time, a form of practical sense which operates without the necessary mediation of conscious thought but which is radically different from the simple application of a set of acontextual abstract rules. In other words, the habitus is a dynamic immanent structure which imperfectly reproduces the social relations of its past in the strategies of members of future generations. It does not induce knee-jerk or mechanical reactions to events but rather instils creative dispositions, bounded by limits imposed by social conditioning while at the same time mediating a whole variety of reactions to what must always in some respects be the unique circumstances in which individuals find themselves.

> Action is not the mere carrying out of a rule, or obedience to a rule. Social agents, in archaic societies as well as in ours, are not automata regulated like clocks, in accordance with laws which they do not understand. (Bourdieu 1991, p9)

Bourdieu claims that in ancient societies there are very few explicit rules, that society is regulated by the reproduction of the habitus within a shared but largely unspoken world view. Ancient societies can operate in this way because they are more culturally homogeneous. We are dealing with community (*Gemeinschaft*) rather than a society or association (*Gesellschaft*). Bourdieu refers to the experience of this unspoken world view as a 'doxa'. Traditional societies have a communal doxa:

> in the extreme case, that is to say, when there is a quasi-perfect correspondence between the objective order and the subjective principles of organisation (as in ancient societies) the natural and social world appears as self-evident. This experience we shall call *doxa*, so as to distinguish it from orthodox and heterodox belief implying awareness and recognition of the possibilities of different or antagonistic beliefs. (p164)

In traditional societies power distribution and social values are relatively uncontested; they are untheorised and so largely unquestionable, forming the second nature of all who live in that community. As Eagleton puts it, paraphrasing Bourdieu, 'what matters in such societies is what "goes without saying" which is determined by tradition; and tradition is always "silent", not least about itself' (Eagleton 1991, p157). The incorporation of the habitus to reproduce communities structured by a common doxa is then one mode of operation of ideology. It reproduces a form of life and its associated dispositions and values in a manner such that they remain unquestioned and unquestionable, stable over many generations and relatively unchanging. Non-conformity would be rare in such a society since all are inculcated by the same habitus and incorporate the same world view. Ethical values are stable and shared by all members of society in respect of their given roles in that society. There is little need for, or possibility of, ethical and meta-ethical speculation. Indeed, Bourdieu claims, there is little need for theory at all in traditional societies. The transmission of the habitus occurs through the experience of practices themselves rather than through the medium of theoretical discourse. Bodily communication performs a much more important function – bodily *hexis* is incorporated *directly* into the individual's dispositions.

> So long as the work of education is not clearly institutionalised as a specific practice . . . the essential part of the *modus operandi* which defines practical mastery is transmitted in practices, in its practical state, without attaining the level of discourse. (Bourdieu 1991, p87)

Our deportment, body language and forms of life are incorporated and reproduced without being theoretically articulated as we are brought up and interpellated into certain communally recognised niches. All of this changes radically in modern society. The previously homogeneous community is fragmented by continuous and rapid change and by the proliferation of disparate practices. The increasing complexity of society and the increasing specialisation seen within it diminish the degree to which everyday practical life can be shared by all members of that society. Each has to find some method of communicating his or her values and dispositions if the society is to continue to function, for the values and dispositions which develop within these relatively autonomous fields of society may be radically different. As values can no longer be inculcated through direct experience of those practices, the spoken and written word together with other methods of mass communication come to mediate between increasingly isolated individuals, each enveloped within, and formed by, his or her own unique place in a set of social fields. Efficiency demands that inculcation can no longer occur simply by practical participation in society without the mediacy of theoretical discourse and formal education systems. Discourse in general and theoretical discourse in particular become the primary mode of operation of ideology (although Bourdieu would not necessarily express the difference in terms of *ideology,* a term which he rejects on the grounds that it both lacks specificity and is often treated as synonymous with false consciousness). But as theory becomes the locus of ideological transmission, discourse has to *codify* practices which were previously experienced through other levels of society. Explicit rules become more and more necessary to maintain social coherence as the doxa is challenged and dispersed.

> It is when the social world loses its character as a natural phenomenon that the question of the natural or conventional character . . . of social facts can be posed. (Bourdieu 1991, p169)

In societies like our own which are in a constant state of flux, which move from crisis to crisis, theory has to take an increasingly active role. It does this by codifying practical experience, reducing it to clear, simple, basic formulae which because of their simplicity and generality are communicable between members of that society. Such formal theory is one way of ensuring at least a minimal degree of communality.

To codify is to come to regulate social practices by formal rules – to *objectify* the previously unspoken doxa in a juridical discourse, to impose a symbolic order. 'Codification is an operation of symbolic ordering, or of the maintenance of the symbolic order' (Bourdieu 1991, p80). As more and more of the society's activities become objectified in this way, the doxa becomes less influential and theory becomes a site of conflict, where experiences of practices clash with, or agree with, the expression given to or denied to them in theoretical practice. The implicit doxa is replaced by an explicit orthodoxy which, because it no longer has the unquestioning consent associated with the doxa, can be challenged by explicit heterodoxies.

A process of increasing codification has engulfed the modern social formation. It is expressed in an understanding of theory as a representation of the 'external world'. As theory becomes more abstract and autonomous it begins to picture itself as a *world apart* from practice, a world of pure thought mirroring that which is external to the mind. (See Richard Rorty (1988) for a more detailed account.) At the same time the dissolution of the doxa and the increasing isolation of the individual lead to the conception of a subjective realm of thought. Thus the poles of objective/subjective reason are born. The ability to communicate and describe the world is put down to a universally shared rationality, an ability to think and calculate directly and consciously about all our actions and values. The reification of this concept as neutral 'reason' could only have happened in a framework where codification has assumed such power and control.

The picture which Bourdieu paints of modern society is one where an explicit logic rules both society as a whole and the individuals who compose it. This formal logic finds the vagaries of everyday practical logic anathema, and imposes its own quasi-juridical definition of reality. It tries to apply its own criteria to habitual behaviour by claiming to excavate a logic or grammar which underpins everyday life, a logic which is not really there but is a fiction of its own theoretical/practical relationship to the dispositions it observes. The form taken by current moral theory is a product of our society's interaction with the world and not an expression of the ontological structure of the world.

Bourdieu's problematic can be seen as adding a Wittgensteinian subtlety to the Frankfurt School's analysis of codified discourse in modern society. Bourdieu places a necessary emphasis on theory's ideological operations, reminding us that theory operates unconsciously in defining our relations to social taxonomies and values. He dismisses the idea of a neutral rationality, deeming it a form of rhetoric. All discourse is imbued with a certain symbolic capital, it expresses certain power relations within society.

The theoretical forms of modern ethics are not set in stone. Those who seek to develop alternatives to society's hierarchical and centralised institutions can recognise and dispute the unconscious extension of systems of formal rationality into the ethical sphere. The values of a radical environmentalism, a movement seeking social change, cannot simply be incorporated in a moral framework which is derived from and supports our destructive society. Both the form and content of this theoretical sphere must change, and critics of contemporary society need to develop alternative moral discourses in which to express their values. What forms any alternative ethical discourses might take is not yet clear. However, a number of possibilities have been mooted, most particularly in the discourse ethics of Habermas and Benhabib and the ethics of care associated with Gilligan's feminist critique of theory. These possibilities are open for discussion.

Some, including 'postmodern' environmentalists such as Jim Cheney, have argued that we need to break with the hegemony of theory itself and return to contextual and less abstract modes of thought and action. Although sympathetic towards the need for contextual discourse, I would argue that we cannot reject theory *en masse;* rather, we need to recontextualise 'theory'. Even if it were possible to revert to an atheoretical 'primitivism', we should not give up the very real freedoms which critical theorising can produce in favour of submersion in an unquestionable communal doxa, however environmentally friendly this might be (Smith 1993). Our feel for the game of life must include a feel for the possibilities of theory itself once it has shed the armour of formalisation and regulation. This theoretical ecdysis has to wake us from the frozen one-dimensionality of current moral philosophy and let us fly once more. When not used as an agent of social control, theory too can weave enchantments and create new insights into the value of our truly magical world.

In summary, the whole project of developing abstract theoretical methodologies to *represent* environmental values, particularly ethical values (so beloved of governmental funding bodies), is problematic. Genuinely oppositional moral values, values which refuse to be reduced to utility or rights, are excluded from the public sphere. For this reason it is important that radical environmentalism does not fall victim to the interminable maze of moral argument which currently passes as the sum total of the moral sphere. Those moral theories, like utilitarianism, which have a practicable intelligibility in terms of a reductive efficiency in modern bureaucratic society would lose this entirely in a society reconstituted and restructured on radically different lines. Any engagement with current moral paradigms may actually be counterproductive in so far as reaching any radically different form of social and environmental relations is concerned. We simply reinforce a form of philosophy which is complicit in the incorporation of environmental values into the present social structure, rather than advocating social change.

NOTE

1. Thanks to Valerie Allen for drawing my attention to Marvell and to Joyce Davidson for her comments.

REFERENCES

Bauman, Z. (1993) *Postmodern Ethics,* Blackwell, Oxford.
Benjamin, W. (1992) 'Theses on the Philosophy of History', in *Illuminations,* Jonathan Cape, London.
Bourdieu, P. (1991) *Outline of a Theory of Practice,* Cambridge University Press, Cambridge. Translation of *Esquisse d'une théorie de la pratique* (1972) Droz, Geneva.
Eagleton, T. (1991) *Ideology: an Introduction,* Verso, London.
MacIntyre, A. (1981) 'A Crisis in Moral Philosophy' in D. Callaghan and H. T. Engelhardt Jr (eds) *The Roots of Ethics: Science, Religion and Values,* Plenum Press, New York.
Rorty, R. (1988) *Philosophy and the Mirror of Nature,* Blackwell, Oxford.
Smith, M. (1991) 'Letting in the Jungle', *Journal of Applied Philosophy,* **8**, 2, 145–54.
Smith, M. (1993) 'Cheney and the Myth of Postmodernism', *Environmental Ethics,* **15**, 3–17.

Section II

EVALUATING
ENVIRONMENTAL VALUES

Introduction

JONATHAN CHASE
Department of Psychology, University of Surrey

The chapters in this section of *Values and the Environment* consider the question of how one can assess people's evaluations of the environment or aspects thereof. This is an important issue and has theoretical, methodological and policy implications. Theoretically, the question gives rise to consideration of the nature of (environmental) values – their stability, structure and origin. Methodologically, it prioritises consideration of the way in which the assessment may affect the expression of any particular value. Politically, it suggests that expressed values need to be treated with caution and not be reified, thus assuming a reality and permanence that is unwarranted. The various chapters in this section contribute to the discussion of these issues.

A number of techniques for assessing people's values regarding environmental issues have emerged in recent years. Foremost among these have been the contingent valuation method (CVM), which has mainly been developed by economists, and attitudinal measures, which derive from psychological research. However, it is clear that different methods of elicitation often, indeed usually, provide very different evaluations of particular environmental issues.

All the chapters in this section examine, albeit by different means, how the values applied to environmental issues may be actively constructed rather than passively retrieved. Thus they represent, in their own ways, a sustained critique of the assumption that one can simply elicit stable values for environmental issues, events or goods. Values relating to the environment are not unproblematic but reflect the operation of complex psychological, social and political processes. For instance, the presentation of the same question in different contexts (eg individual freedom versus maintaining the environment) can even lead to apparent reversals of preference by respondents. The idea of the 'true' response, with its implicit appeal to a context-free and therefore context-spanning reality, becomes redundant.

Similarly, the chapters in this section suggest that the idea that some values can be labelled 'environmental' and others cannot is also untenable. This assumption reflects a recapitulation of lay discourse by which 'environment', 'environmental values', 'environmental behaviour' and other such terms are treated as unproblematic. Nevertheless, one implication of Burningham's paper is that the processes through which people and the media represent particular

Values and the Environment: A Social Science Perspective, Edited by Yvonne Guerrier, Nicholas Alexander, Jonathan Chase and Martin O'Brien. © 1995 John Wiley & Sons Ltd.

values as 'environmental' (an aspect of what might be termed the discursive construction of the environment) are worthy of research. However, researchers need to be aware of the problems surrounding such an uncritical approach to the relationship between values and behaviour regarding the environment.

Chapter 5 considers some of the processes which may be involved in the generation of values regarding environmental issues. It suggests that identity processes may be important factors in this process. Values are associated with identities and may, therefore, be a function of the identity salient at the time. It is argued that environmental issues are easily seen as possessing conflictual aspects and that identity is important in determining whether or not this is so. Particular identities tend to emphasise different values and to place more or less weight on other people's interests. This is seen as especially relevant to the temporal dimension of environmental issues. A better understanding of the process by which values become applied to instances would help policy makers predict the public's (or publics') reaction to potential and new environmental issues.

Chapter 6 focuses on the use of CVM, which is one of the main techniques used by economists and others to provide monetary values for environmental goods. This chapter provides a review of the use of CVM and of various criticisms of it. Clark suggests that problems with CVM cannot be resolved simply by developing more sensitive or otherwise better survey instruments. She argues that environmental values are more complex than is assumed by the use of CVM, and that therefore assessment methodologies have to reflect this complexity. This chapter provides a strong theoretical critique of CVM.

Burningham in Chapter 7 draws on 'discourse analysis', a set of theoretical and methodological approaches based on linguistics and psychology, to critique the assumption that one can simply uncover people's values. Discourse analysis problematises the assumed linkage between overt expressions of beliefs, attitudes and values and the respective covert mental states; discourse analysis emphasises the rules governing expression and account giving. Burningham uses this theoretical perspective to examine the different discourses provided by different groups regarding a road 'improvement' scheme. She shows how both pro- and anti-road groups seek to legitimise their own claims and undermine those of their opponents. Both groups, for instance, lay claim to being pro-environmental and represent their own position as reflecting 'objective' truths. Burningham's research helps to elucidate the rhetorical strategies by which particular groups act socially to construct environmental issues.

Thus, these chapters represent a sustained critique of the dominant approaches to deriving values for environmental issues and goods. Although the theoretical perspectives informing these criticisms vary between the respective authors (and may not be reconcilable one with another), there are a number of common threads which run through all three chapters. These include the following.

Values are not seen as being necessarily stable. There is no underlying utility function, for instance, from which preferences and values are retrieved in a consistent and reliable manner. Instead, the expression of values (both representationally in discourse and in action) varies depending on context.

Values are complex rather than simple; thus the imposition of a single dimension, be it monetary or whatever, is inappropriate and represents a Procrustean strategy. The relationship between values and the way in which value conflicts at the individual level are dealt with need to be explored.

5 Environmental Values and Social Psychology: A European Common Market or Commons' Dilemma?

JONATHAN CHASE AND IOANNIS S. PANAGOPOULOS
Department of Psychology, University of Surrey

INTRODUCTION

The focus of this chapter is on the factors affecting how people value environmental issues and the potential for such issues to become sources of conflict. This question will be addressed from the perspective of social psychology and will, therefore, mainly draw on social psychological theories. In particular, it will draw on theories of identity and of social dilemmas. These theories will be used to conceptualise the dynamic between representation, resources and regions which lies at the heart of environmental issues.

The chapter will start by discussing how environmental issues may be characterised according to abstract rather than substantive criteria (eg in terms of the temporal and spatial distribution of costs and benefits rather than in terms of threats to biodiversity or atmospheric change). Secondly, the form of these distributions will be considered according to the theory of social dilemmas. This theory focuses on conflicts of interest. Thirdly, it will be argued that values are less stable than is assumed by most current approaches. The instability of values requires that the conditions determining their generation be understood. This chapter suggests that identification processes are important factors in the process of valuation. It is argued that certain identities are likely to be associated with environmental issues in predictable ways, eg in regard to the importance given to future generations. Particular types of identification will also be more or less likely to result in competitive rather than co-operative perceptions of environmental issues. Finally, some conclusions are briefly presented.

CHARACTERISING ENVIRONMENTAL ISSUES

Environmental issues are complex and can be characterised in different ways. One could classify them as relating to transport or to energy use, as affecting air, soil or water, or according to their physical properties. Alternatively, one

Values and the Environment: A Social Science Perspective, Edited by Yvonne Guerrier, Nicholas Alexander, Jonathan Chase and Martin O'Brien. © 1995 John Wiley & Sons Ltd.

can examine the distribution of various factors related to the issue. For example, what is the distribution of causal factors – whose actions are causing what effects? What is the distribution of effects – who is being affected by the target phenomenon? How are these effects valued and how are the negatively and positively valued effects or consequences distributed? It is this approach that will be adopted for the purposes of this chapter.

The distribution of costs and benefits (not necessarily strictly monetary ones) associated with various environmental issues may vary along both temporal and spatial, and therefore social, dimensions. For instance, over-fishing may provide an immediate, local gain for some people but constitute a long-term, global loss to many others as that resource fails to replenish itself. The generation of electricity by nuclear power provides an answer to present energy shortages for some groups, but leaves future generations to deal with clean-up and decommissioning costs and causes other groups to suffer increases in exposure to radiation.

The complexity of these distributions of causes and consequences, which can be seen to cross both national and generational boundaries, means that environmental issues are surrounded by uncertainty. The more distant temporally the events or outcomes, the greater the uncertainty associated with them. However, there is likely to be uncertainty about the following aspects of present as well as future environmental problems:

1. Uncertainty regarding the expected outcomes. (Will global warming give Siberia a temperate climate? How does acid rain affect trees? Will a technological 'solution' be developed within a given timespan? Does increased CO_2 mean larger harvests?)
2. Uncertainty regarding the values present. (Will food be cheaper but 'tropical' illnesses more common? Who stands to benefit or to suffer? When? How does one trade off economic versus aesthetic values, spiritual versus materialistic values?)
3. Uncertainty regarding the decisions to be made. (Should fishing be limited and, if so, by how much and to whom? Do I use lead-free petrol or not?)

This uncertainty is present in regard to both scientific and lay understanding of environmental issues and means that it is very difficult (or impossible) for researchers to provide definitive characterisations of environmental issues. It is therefore important that researchers provide explicit definitions of how they are characterising and valuing any particular environmental issue. To take my own advice, this chapter will use the theory of social dilemmas to consider the potential conflicts between self- and other- interest present in environmental issues and behaviours. This theory is discussed in the next section.

SOCIAL DILEMMAS AND ENVIRONMENTAL ISSUES

Social dilemmas are a form of collective action, that is, the outcome is a consequence of more than one actor (person, organism, business, organisation, state, etc). In the simplest form of social dilemma, each individual faces a binary choice, one choice gives him or her a better outcome than the other. However, if everyone chooses the better or dominant choice then all are worse off than if everyone had made the other, dominated choice (Dawes 1980). In other words, one choice is rational for the individual as an individual, while the other choice is rational for the individual as a member of the collective or group. This conflict of interest is often described as being between the individual and the group, but is better described as a conflict between the individual as an individual and the individual as a member of the group. It should be noted that the social dilemma is defined by the structure of choices and outcomes. The way in which people value different outcomes therefore actively constructs whether or not the situation has social dilemmatic properties for them.

Social dilemmas may be present between the interests of individual actors (at various levels – nation state, commercial organisation, age cohort, geographical area, individual) as individuals and as members of collectivities (UN, EU, nation state, commercial union, family, etc). This conflict of interests may also exist within an individual actor (eg between present and future self).

Environmental problems have been considered to have the characteristics of social dilemmas (Dawes 1980; Colman 1992; Sandler 1992; Chase 1992). This may explain the discrepancy between the high levels of self-reported environmental concern and of pro-environmental attitudes (eg Krause (1993) found that about 75 per cent of respondents described themselves as environmentalists) and the low levels of pro-environmental behaviour. For instance, when asked about public transport many people are in favour of it, but say they themselves could not use it. This is exactly what one would expect if environmental action is social dilemmatic. Even though each person desires a common outcome, the individually rational action is to do nothing to provide for that outcome. Some environmental issues have been researched using the theory of social dilemmas (eg resource use by Rutte et al 1987; O'Connor and Tindall 1990).

Vlek and Keren (1992) suggest that environmental problems may be characterised by other types of dilemma. These are benefit–risk, temporal, spatial and social dilemmas. Any specific environmental problem may show a combination of these four types depending on the peculiarities of that instance (for instance, nuclear power presents a greater temporal dilemma than does acid rain). This suggests that environmental issues may be conceptualised as having:

1. a multi-dimensional utility structure;
2. a distribution of outcomes across space;
3. a distribution of outcomes across time.

Consequently, environmental issues involve value/preference conflicts. These value conflicts are within individuals (eg between different values at the same time or between present and future selves), between individuals, between individuals and groups, and between groups and groups.

An example may show this more clearly. If I am concerned about global warming I could buy a catalytic converter for my car to cut down the discharge of exhaust gases which contribute (fractionally) to global warming. However, the benefit I gain from this is minute and equal to the benefit which every other individual also gains from my action, while the cost is solely to myself. Additionally, my contribution is so slight that unless many (millions) of other people also act global warming will still occur. Finally, if everyone else does act so as to reduce global warming but I do not then I will still benefit, global warming will be significantly reduced, but I will not incur the costs. According to this type of individual rationality, which is the standard assumption of modern economics, short-term selfishness will dominate long-term altruism.

VALUES AND THE ENVIRONMENT

Environmental issues may involve different values. The initial question which this section addresses is what values are likely to be present when people consider environmental issues? The above discussion suggests that both selfish values and collective ones may be present and that an economic dimension is particularly likely to be relevant.

The main approaches to explaining environmental behaviours and actions have focused on knowledge, attitude and/or personality factors. That is, attempts are made to explain pro-environmental behaviour in terms of a person's attitudes towards, and/or knowledge about, a specific environmental target or in terms of the presence of a specific 'green' personality (eg Christianson and Arcury 1992; Lansana 1992; Herrera 1992). The attitudinal and knowledge approach has led to the suggestion that there is an identifiable constellation of attitudes and knowledge typifying those people who engage in 'pro-environmental action'. An example of this is the 'new environmental paradigm' (Van Liere and Dunlap 1978; DeHaven-Smith 1988).

For instance, a large amount of research has been done on the effects of the Chernobyl incident on public opinion using an attitudinal approach (eg Midden and Verplanken 1990; Peters et al 1990). However, even when data were gathered in more than country (eg Eiser et al 1990) attitudes are simply compared between nations, with no analysis (nor indeed relevant data collected) of how identification processes, in this case national identities, might affect the representation of environmental issues, the attribution of responsibility for their causes and remedies, the perception of costs and benefits, etc.

West Germans were found to consider the probability of a Chernobyl-type accident to be much less in West Germany than in the USSR (Peters et al 1990). Given that most accidents are the product of human error, the superiority of West German technology does not explain this result. While perceived differences in regulatory practices might also explain this result, group processes may be present. It suggests that the risk from reactor accidents is seen as coming from an out-group rather than from the in-group.

From an economic perspective (which, understandably, prioritises monetary values) the question of how to identify the value of various aspects of the environment has led to the development of methods such as the contingent valuation method (CVM). These approaches again assume that values are relatively stable, unitary and individual. However, recent work in social and cognitive psychology suggests that values (and other cognitive phenomena such as attitudes) may be more dynamic and complex. Kahneman et al (1993) found that the values derived from a 'willingness-to-pay' methodology were strongly correlated with standard attitudinal measures about the same issues. This suggests that 'willingness to pay' is essentially a measure of attitude rather than of economic valuation and that, therefore, the research suggesting that attitudes are unstable may be safely generalised to CVM.

One strand of this work on attitudes emphasises the way in which values are talked about and constructed in conversation. Another approach focuses on the active generation of values in response to perceived contextual demands (which might include conversational ones) and cues, rather than their passive retrieval from memory as assumed by economic models. Both of these approaches would tend to suggest that the same issue or event might be valued in different ways by the same person at different times, and that it is mistaken to argue that one or other value represents the 'true' value. This is so regardless of whether or not the same metric or value system is present. In other words, if the value system is monetary (a classic economic valuation) then someone may suggest different monetary values depending on the context (value elicitation technique, identity salience, current events etc). Alternatively, someone may emphasise and assess monetary values in one context and aesthetic ones in another. In this context a person might be both an environmentalist and a motorist; however, awareness of, and identification with, these group memberships vary. If people identify themselves as members of an environmental group – as environmentalists – then they may place more value on environmental features (eg Twyford Down, a notorious beauty spot that became a battlefield between various anti-road groups and the developers when the M3 extension was carved through it) than if they were conscious of being motorists. When driving one wants good roads, when picnicking good views. The individual may feel no need to resolve this inconsistency unless directly confronted with it.

As the example suggests, people's valuations of environmental issues, events, etc may be a function of the identity salient at the time. A standard assumption in modern social psychology is that people engage in identification processes which result in their internalisation of perceived membership of various social groups (eg Tajfel 1981; Oakes et al 1994). This process includes the development of beliefs about the defining characteristics, attributes and values of group members. Differences between groups are emphasised, especially on important dimensions. Behaviour may also be affected so that discrimination in favour of the in-group and against the out-group occurs. The values that are active are a function of the identification current at the time. For instance,

Feather (1994) found that strong collectivist values were related to more in-group favouritism between national groups. As in-group identification varies one would expect the importance of collective values to co-vary.

Many environmental issues relate to groups or collectivities; these may include families, commercial organisations, people with common occupations (eg fishers, farmers) or interests (eg gardeners, cyclists, fox hunters), and nations. Does the application to groups of the same logic as that underpinning individual rationality lead to the same result in these cases? Given that the effects of environmental change consequent on the actions of (some) humans often involve the differential distribution of costs and benefits across social groups, the potential for increasingly competitive interaction between groups must be considered to be present, and this competition may possibly be greater than that found when purely individualistic interaction occurs. An example of this group level of conflict can be seen at present in the case of Slovakian plans to divert the Danube which are causing concern in Hungary and conflict between the two countries. Acid rain is another example. The UK 'exports' various air pollutants to Scandinavian and North European countries. Those countries have to bear some of the cost of UK energy production but gain no benefit. At an individual level of analysis, the judged benefits and perceived voluntariness of exposure to risk have been found to be related to both risk estimates and risk tolerance (Baird 1986). In regard to acid rain, these risk perception factors would map onto national identities with some countries being exposed to involuntary costs by other countries. One would expect that nationals of the former countries would rate the risks as greater than would the nationals of the latter countries.

Values may be related to identities in two ways. First, particular values may be associated with particular identities. So, for instance, a communist identity is necessarily associated with collective rather than individualistic values; religious identities are associated with spiritual values, business identities with financial values and so on. As one may be both a businessperson and a religious one, the relative salience of the two identities may determine which potential value system (economic or spiritual) is activated and applied to the instance.

Secondly, identity salience may determine the perception of the distribution of costs and benefits, ie it may determine how that value is applied to the specific case. For instance, a salient identity as a parent may result in more value being placed on future outcomes and the values associated with them (positive or negative) than would be the case if such a parental identity were absent. An identity of British may lead one to unitise costs and benefits associated with the unit 'Britain' by including outcomes for other people identified as 'British' in one aggregate outcome, and so on.

To summarise the argument proposed here:

1. Different values may be applied to environmental issues.
2. Values are associated with social identities. Therefore:

3. The way in which people value environmental issues will be a function of the relative salience of social identities.

Thus the question asked at the beginning of this section therefore becomes, what identification will be present when people consider environmental issues? This question may be further refined by considering whether particular environmental issues are associated with particular identities.

One type of identity which may be salient when environmental issues come to someone's attention is that of the 'environmentalist'. This may be formalised through membership of an environmental organisation such as Greenpeace. Greenpeace, Friends of the Earth and other environmental groups are international organisations. Whether relating to membership of an organisation or not, environmental identities tend to emphasise collective and long-term interests over individual and short-term ones. They also tend to focus on non-human interests (either non-human species or abstract entities such as 'Gaia').

Conversely, environmental issues are also economic ones and relate to access and control of resources. These economic values are related to ownership and possession. Although economic values are often represented as individual they are not necessarily so. Group memberships and identities, such as nationality, can be important in regard to legal and economic 'rights'. Such identities emphasise differences between groups and individuals. The immediate costs and benefits of various activities which affect the environment are rarely evenly or equitably spread. For instance, the benefits of burning coal in power stations go to the owners (and possibly the consumers) of those power stations, while the costs of acid rain are typically exported to another state. This would be reversed in regard to the costs of cleaning up such coal-burning power stations. These cost–benefit distributions often map onto objectively defined social groups, such as classes, ethnic groups and especially nations.

National identities may also be relevant to environmental issues in other ways. Environmental change may affect national identity. Melman (1991) analysed historical novels and conventional histories regarding the development of an 'English identity'. He identified three factors, of which one was the construction of a national geography defining England. This representation provided geographical boundaries and a particular content – 'an emerald set in a silver sea'. The representation of the English climate as temperate resonates with that of the English character as tolerant. (Perhaps both of these representations have taken a battering in recent years.) Thus, representations of national character or identity may be intimately bound up with representations and experience of the physical environment. The experience and representation of the physical environment provide a figurative nucleus around which a social representation of a national identity can form. (Incidentally, this also shows the importance of analysing the specific content of an identity representation as well as the general process of identification.)

Referring back to the question asked at the start of the section, this analysis suggests that environmental issues which cause dramatic and observable effects on the landscape are likely to activate national identities, especially when those features are central to the representation of the nation. So, acid rain as an issue in the UK will not be associated with national identity, but it will be in Germany where the forests are a core element in German national identity.

The presence of a salient national identity may also determine how outcomes are aggregated and whether the costs or benefits of an action are salient. For instance, for Norwegians the immediate costs of stopping whaling may be more salient than the benefits, while the reverse could be true of members of non-whaling nations. Even when an 'environmental identity' is present (eg membership of Greenpeace) and common across national boundaries, it may be fragile and easily overwhelmed by national identity once a 'conflict between nations' frame is present. For instance, worker members of international socialist groups before the First World War tended to respond to the conflict at a national rather than international level of identification.

In regard to behaviour in social dilemmas, the presence of a common social identity has been found to reduce competition and increase co-operation (Kramer and Brewer 1984; Brewer and Kramer 1986). However, this is true only when all parties to the interaction share the same common identity. The presence of two groups ('us' and 'them') leads to more intense conflict than when individual identities are salient. Thus, the 'nationalising' of environmental issues would be expected to increase competitive and decrease co-operative interaction. This may well lead to less effective management of the issue (eg failure to restrain fishing, or inability to limit exhaust and other emissions). As the problem became more acute conflict would intensify and a spiral of decreasing co-operation ensue.

The importance of social and, therefore, identity factors can also be seen in the way in which environmental issues are constructed and given meaning by media and special interest group actions. An analysis at the individual level fails to investigate this social construction of environmental issues, a process which is prior to environmental behaviour and may contribute to the type (if any) of behaviour required on the part of the individual. Thus, a purely individualistic approach is inadequate in regard to environmental issues.

However, it needs to be emphasised that identities are not static but dynamic; new identities emerge and old ones change. The structure and processes of specific global environmental change (GEC) events may actively construct social groups through the creation of a common fate for a (potential) set of individuals. So, for instance, the threat to Oxley's Wood in south east London from the planned South East London River Crossing led to the formation of protest groups, the members of which were quite disparate (eg pagans, bird watchers, parents, Friends of the Earth, etc). This is as well as individuals' membership of social groups determining their positioning within the dilemma structure, ie their particular set of costs and benefits.

Environmental change necessarily affects the physical environment in which we live. Research on place identity (eg Proshansky et al 1983) and on social and individual memory (Fentress and Wickham 1992) shows how one's physical surroundings contribute to one's sense of identity. Therefore, a changed physical environment may change the self-concept. (This may be especially true of 'natural' environments; Sebba (1991) found that almost all adult subjects identified the most important place in their childhood with the outdoors.) Unstable environments will retain fewer cues to past cognitive, affective and behavioural states. The effects of global environmental change may radically alter the local climate, the ecology and even the geography as patterns of erosion and deposition shift. Thus, GEC may result in changes in the self-concept. Decreasing continuity in the physical environment may threaten continuity in the self.

IDENTITY, TEMPORAL DISCOUNTING AND FUTURE GENERATIONS

The importance of an identity perspective on the valuation of the environment can be further illustrated through a consideration of how future aspects – the temporal dimension – have been considered and how they could be considered. The two main approaches which have been taken regarding the discounting of the value of future outcomes are economic and decision analytic. These temporal aspects of behaviour have been examined largely at the level of the individual.

In the case of economics this involves an assumption of individual rational selfish utility maximisation, ie an idealised model of the person is adopted. All people are defined in the same manner and data tends to be collected at the macro level although surveys are sometimes used (eg in 'willingness-to-pay' studies). Decision theory is more informed by cognitive psychology and, therefore, is more likely to include deviations from full rationality. For instance, different presentations of the same problem ('framing') may be compared. An example of a framing effect is the difference in amount required to compensate for the loss of a good, amenity or asset compared to the amount one would be prepared to pay for that good, etc. Other framing effects may also be relevant. Many studies have found that people tend to discount the value of future outcomes against that of present ones, which may, partly, be explainable in terms of different frames being accessed in each of the two cases. Vining (1987) found that people's environmental decisions were sensitive to different decision contexts and to different writing styles, which shows that framing effects can occur in environmental decision making.

The complex nature of environmental problems suggests that it is unlikely that either of these approaches will be successful in fully explaining the determinants of people's behaviour in regard to environmental issues. The reason for this is the failure of either approach to include any consideration of the social factors involved in the process of discounting. One way to understand discounting considers future outcomes as progressively more uncertain, ie discounting is a function of uncertainty. Bjorkman (1984) argues that future events are more ambiguous (as one has less knowledge about them) and elicit less concern (through low involvement). Perceptions of control and, perhaps, self-efficacy may also be lower for distant than for close future events. People may estimate the probability of outcomes by constructing scenarios of future states and use an assessment of the ease of scenario construction as an index of probability. These scenarios may include causal models explaining the presence or absence of the target state. Most of this work looks at how the individual discounts his or her own future outcomes, although Kunreuther et al (1990) did find that resistance to a nuclear waste repository in Nevada was related to the perceived risk to future generations and the level of trust placed in the relevant governmental agencies.

All these factors are, in part, a function of social processes, such as the way in which the future (and past) are represented in a culture, the images and discourses about the environment, group and social identity processes etc. The inclusion of this level of analysis offers better hope for inducing large-scale change in behaviour than would a solely individualistic one.

Given that the discounting function is likely to be actively constructed rather than passively retrieved from memory, it is vital to consider what factors affect discount function construction. For instance, the role of identification processes in the discounting of temporally and spatially distant outcomes needs to be explored.

Even though environmental issues often include very long timespans beyond the individual's life, the temporal dimension of environmental problems has seen less attention than has the spatial dimension. Present actions may have an impact on future generations. This question will now be considered.

Strictly future generations (ie excluding any living people including the foetal) only exist as a potential and as representations in the minds of actual people and in cultural artifacts. Thus, interaction with future generations is necessarily mediated by representational (social and cognitive) processes and does not include actual interaction. The inclusion of the impact of one's own actions on future generations (a less narrowly 'selfish' utility function, cf Margolis 1981) in the decision-making process would imply some degree of identification with those 'people'. Therefore, processes of both representation and identification are fundamental to understanding how people consider future generations.

The following questions may be considered:

1. How do people represent the future and future generations and what is the content of these representations? For instance, are children taken to represent the future?
2. Are different representations associated with different behaviours, intentions, attitudes etc?
3. Do members of different groups have different representations of the future?

Some research suggests that a person's time perspective is a function of their age, education and health as well as their domain of action. Fischer et al (1991) found that risks to the environment were more commonly perceived by students and females than by males and older people. By and large, willingness to pay was greater to reduce personal risks than to reduce risks to the environment or to people in general. What is the relationship between discounting and age? Do the young give greater weight to future outcomes than the elderly? Research on saving suggests that older people (60+ years) save more than purely selfish utility maximisation would predict. However, very few respondents actually responded that saving was intended as a bequest to subsequent generations (Projector and Weiss 1966).

CONCLUSION

This chapter has argued that the values accessed regarding any specific environmental issue are not stable and will change from context to context. It is argued that these values are partly a function of the identity salient in any particular context. Selfish values will tend to predominate over more altruistic ones. However, selfishness may operate at either the individual (individual egoism) or group (group egoism) level. Identity processes have also been shown to be important in regard to the temporal dimension of environmental issues, especially in regard to how people consider future generations' interests.

It has been suggested that nationality is an especially salient and highly accessible identity which will be cued by stimuli about environmental issues (eg 'What do you think about whaling?'). The framing of environmental issues in terms of national interests will tend to lead to perceptions of the environmental issue in conflictual terms and with the properties of a social dilemma. This may lead to the development of an increasingly competitive interaction. A number of environmental issues have been identified which seem particularly prone to this type of framing.

The paradox of environmental issues is that they require collective action to manage or solve them but tend to elicit an intermediate level of identification which exacerbates conflict by pitching it at an inter-group level. Collective action does occur but, given the level of the framing – the world made up of competing groups (often nation states) rather than of one superordinate common humanity – it is the action of competing collectivities rather than of one co-operating group.

REFERENCES

Baird, B. N. (1986) 'Tolerance for Environmental Health Risks: the Influence of Knowledge, Benefits, Voluntariness, and Environmental Attitudes', *Risk Analysis*, **6**, 4, 425–35.

Bjorkman, M. (1984) 'Decision Making, Risk Taking and Psychological Time: Review of Empirical Findings and Psychological Theory', *Scandinavian Journal of Psychology*, **25**, 31–49.

Brewer, M. B. and Kramer, R. M. (1986) 'Choice Behaviour in Social Dilemmas: Effects of Social Identity, Group Size, and Decision Framing', *Journal of Personality and Social Psychology*, **50**, 543–9.

Chase, J. (1992) 'The Self and Collective Action: Dilemmatic Identities?', in G. M. Breakwell (ed) *The Social Psychology of Identity and the Self Concept*, Academic Press, London.

Christianson, E. H. and Arcury, T. A. (1992) 'Regional Diversity in Environmental Attitudes, Knowledge and Policy: The Kentucky River Authority', *Human Organization*, **51**, 99–108.

Colman, A. (1992) *Game Theory and Experimental Games: The Study of Strategic Interaction*, Pergamon Press, Oxford.

Dawes, R. M. (1980) 'Social Dilemmas', *Annual Review of Psychology*, **31**, 169–93.

DeHaven-Smith, L. (1988) 'Environmental Belief Systems: Public Opinion on Land Use Regulation in Florida', *Environment and Behavior*, **20**, 276–99.

Eiser, J. R., Hannover, B., Mann, L., Morin, M. and van der Pligt, J. (1990) 'Nuclear Attitudes After Chernobyl: a Cross-national Study', *Journal of Environmental Psychology*, **10**, 101–10.

Feather, N. T. (1994) 'Values, National Identification and Favouritism Towards the In-group', *British Journal of Social Psychology*, **33**, 467–76.

Fentress, J. and Wickham, C. (1992) *Social Memory: New Perspectives on the Past*, Blackwell, Oxford.

Fischer, G. W., Morgan, M. G., Fischhoff, B., Nair, I. and Lave, L. B. (1991), 'What Risks are People Concerned About?', *Risk Analysis*, **11**, 2, 303–14.

Herrera, M. (1992) 'Environmentalism and Political Participation: Toward a New System of Social Beliefs and Values?', *Journal of Applied Social Psychology*, **22**, 8, 657–76.

Kahneman, D., Ritov, I., Jacowitz, K. E. and Grant, P. (1993) 'Stated Willingness to Pay for Public Goods: a Psychological Perspective', *Psychological Science*, **4**, 5, 310–15.

Kramer, R. M. and Brewer, M. B. (1984) 'Effects of Group Identity on Resource Use in a Simulated Commons' Dilemma', *Journal of Personality and Social Psychology*, **46**, 5, 1044–57.

Krause, D. (1993) 'Environmental Consciousness: an Empirical Study', *Environment and Behavior*, **25**, 1, 126–42.

Kunreuther, H., Easterling, D., Desvousges, W. and Slovic, P. (1990) 'Public Attitudes Toward Siting a High-level Nuclear Waste Repository in Nevada', *Risk Analysis*, **10**, 4, 469–84.

Lansana, F. M. (1992) 'Distinguishing Potential Recyclers from Nonrecyclers: a Basis for Developing Recycling Strategies', *Journal of Environmental Education*, **23**, 2, 16–23.

Margolis, H. (1981) 'A New Model of Rational Choice', *Ethics*, **91**, 265–79.

Melman, B. (1991) 'Claiming the Nation's Past: the Invention of an Anglo-Saxon Tradition', *Journal of Contemporary History*, **26**, 575–95.

Midden, C. J. H. and Verplanken, B. (1990) 'The Stability of Nuclear Attitudes after Chernobyl', *Journal of Environmental Psychology*, **10**, 111–19.

Oakes, P. J., Haslam, S. A. and Turner, J. C. (1994) *Stereotyping and Social Reality*, Blackwell, Oxford.

O'Connor, B. P. and Tindall, D. B. (1990) 'Attributions and Behaviour in a Commons' Dilemma', *Journal of Psychology*, **124**, 5, 485–94.

Peters, H. P., Albrecht, G., Herren, L. and Stegelmann, H. U. (1990) '"Chernobyl" and the Nuclear Power Issue in West German Public Opinion', *Journal of Environmental Psychology*, **10**, 121–34.

Projector, D. S. and Weiss, G. S. (1966) *Survey of Financial Characteristics of Consumers*, Board of Governors of the Federal Reserve System, Washington, D C.

Proshansky, H. M., Fabian, A. K. and Kaminoff, R. (1983) 'Place-identity: Physical World Socialization of the Self', *Journal of Environmental Psychology*, **3**, 57–83.

Rutte, C. G., Wilke, H. A. and Messick, D. M. (1987) 'Scarcity or Abundance Caused by People or the Environment as Determinants of Behaviour in the Resource Dilemma', *Journal of Experimental Social Psychology*, **23**, 3, 208–16.

Sandler, T. (1992) *Collective action: Theory and applications,* Harvester Wheatsheaf, New York.

Sebba, R. (1991) 'The Landscapes of Childhood: The Reflection of Childhood's Environment in Adult Memories and in Children's Attitudes', *Environment and Behavior*, **23**, 395–422.

Tajfel, H. (1981) *Human Groups and Social Categories,* Cambridge University Press, Cambridge.

Van Liere, K. D. and Dunlap, R. E. (1978) 'Moral Norms and Environmental Behavior: An Application of Schwartz's Norm-activation Model to Yard Burning', *Journal of Applied Social Psychology*, **8**, 174–88.

Vining, J. (1987) 'Environmental Decisions: the Interaction of Emotions, Information, and Decision Context', *Journal of Environmental Psychology*, **7**, 1, 13–30.

Vlek, C. and Keren, G. (1992) 'Behavioral Decison Theory and Environmental Risk Management: Assessment and Resolution of Four "Survival" Dilemmas', *Acta Psychologica*, **80**, 249–78.

6 Corncrakes and Cornflakes: The Question of Valuing Nature

JUDY CLARK
Department of Geography, University College, London

The idea of putting a price on nature, on living things and the places they inhabit, arouses strong feelings. On the one hand, placing corncrakes on the same footing as cornflakes is rejected as ethically repellant; on the other, bringing nature within the calculus of economics is asserted to be the best way of ensuring its protection. Such passions might not matter were they confined to academic circles, but as environmental economics has come increasingly to play a role in public decision making (Corkindale 1993; Tunstall and Coker 1992; Department of the Environment 1991) so the question of monetary valuation has come to assume a correspondingly greater significance. And there is much at stake, for debates about monetary valuation reflect deeper conflicts over the construction of environmental choices and which, or whose, criteria should prevail when it comes to making decisions about wildlife and natural habitats.

The purpose here, however, is not to arbitrate between competing interests. In this chapter I consider the claims of monetary valuation to make a valid and legitimate contribution to policy and project decisions concerning the natural world, focusing on the method most frequently used to derive valuations for nature, the contingent valuation method (CVM). At heart, much of the debate about monetary valuation is actually concerned with whether cash values should be assigned to environmental resources, and I argue that in so far as *claims* about the legitimacy of pricing nature reside in *claims* concerning the validity of this method of doing so, debate remains largely at the level of conjecture. The cases of proponents and critics alike rest on untested premises concerning how people actually come to value nature and what they actually value in it. But while uncertainty as to what CVM might measure betokens caution, that in itself does not negate the principle of monetary valuation, and there are additional reasons for misgivings concerning CVM.

CBA AND CVM

Environmental economists using the framework of neoclassical economic theory offer cost–benefit analysis (CBA) as the tool for prescribing, or at least con-

Values and the Environment: A Social Science Perspective, Edited by Yvonne Guerrier, Nicholas Alexander, Jonathan Chase and Martin O'Brien. © 1995 John Wiley & Sons Ltd.

tributing to, choices involving environmental resources. CBA is based on the concept of economic efficiency. Economic developments should make people better off without making any one worse off (in technical terms a Pareto test) and so should only proceed if benefits exceed costs. Conventional CBA compares costs and benefits as measured in market prices, and the argument is that if environmental 'goods and services' are valued using the same measuring rod – money – and added into CBA, then the environment will be properly accounted for in decision making (Pearce et al 1989; Pearce 1991). But for many environmental resources no real markets exist. So economists have had to develop alternative techniques for assessing their monetary value, and of these CVM is the most generally applicable. CVM has been used to value many other environmental resources besides species and habitats; for example recreational benefits, the scenic attractions of landscape, and air and water quality. Other methods besides CVM have been used to value environmental resources (principally hedonic pricing and the travel cost method), but CVM is frequently the only technique available. See Pearce and Turner (1990) for details.

In essence, CVM works like this. A scenario is devised, setting up a hypothetical market for the environmental good or service in question. A sample of the affected population is then surveyed to find out what each respondent would be willing to pay for it. Finally, the individual results are aggregated to give total willingness pay; that is, an estimate of monetary value. For details see Pearce and Turner 1990; Wilks 1990; Mitchell and Carson 1989.

CVM has probably been most widely applied in the USA where its status is now such that its mandatory use in assessing compensation (in cases of damage concerning natural resources covered by the Oil Pollution Act of 1990) is being considered by the government (NOAA 1993). In Britain, using CVM to calculate the value of environmental benefits is a more recent development. In the case of nature, applications include deriving valuations for Sites of Special Scientific Interest (SSSI) in the North East (Mitchell et al 1988), the Yorkshire Dales and Norfolk Broads National Parks (Bateman 1993), and the South Downs and Somerset Levels Environmentally Sensitive Areas (Willis et al 1993).

The validity of CVM for valuing nature rests on several assumptions, the first of which is fundamental to neoclassical economic theory and relates to how individuals make choices. Neoclassical economic theory subscribes to what Cole et al (1991) term 'the subjective preference theory of value', which takes as its starting point 'the individual endowed with tastes and talents and who calculates actions so as to maximise personal welfare or utility' (p6). That is, nothing but the tastes of this hypothetical 'rational economic man' determine his preferences for one commodity over another, and these preferences give rise to specific consumption choices. However, tastes themselves are of no concern to the neoclassical economist. Preferences are treated as given; all that matters about them is that they dictate which consumption choices will give the greatest personal satisfaction. The total demand for any commodity is then the sum of these preferences, and allowing individual preferences free rein leads to the most 'efficient' use of resources by society as a whole. Bound by the premise of rampant 'methodological individualism' (Jacobs 1994), society thus becomes nothing more than an aggregation of individuals acting independently of one another.

Preferences concerning the environment can only be formulated in the same way as preferences for cornflakes or Weetabix if the environment is presented as a set of commodities like any others, which means that it has to be partitioned into specific goods and services (Jacobs 1994). So for the purposes of valuation it also has to be assumed that nature can be satisfactorily divided into discrete sites and species. A preference for a particular bit of nature, say the corncrake, should then be satisfied if the corncrake is protected from harm, and the greater the satisfaction gained from this outcome the more an individual should be willing to pay for it. The aggregated preferences of individuals should thus identify the extent to which society is prepared to protect the corncrake.

Sagoff (1988a; 1988b; 1993), an environmental philosopher and prominent critic of monetary valuation, is scathing when it comes to the notion that environmental choices can be equated with consumer-type preferences. He argues that willingness to pay measures nothing more than that because in confounding the consumer with the citizen CVM commits a classic 'category mistake'. Consumers' preferences are constituted quite differently from citizens' environmental choices, which are based not on the satisfaction of individual wants but on ethical and moral concerns. Sagoff cites protest bids (refusal to play the CVM game) as evidence that people themselves recognise the illegitimacy of pseudo-market valuations as a means of deciding which bits of nature to protect or how much pollution is tolerable. In his view, the environment should remain a field of political negotiation. Environmental choices are matters of the common or public interest of the community as a whole rather than simply the private interests of individuals, and their resolution is properly the province of the political system through which citizens determine the environmental goals to which the community aspires. Should the conservation of the corncrake, for example, be deemed a worthy goal, then it should be pursued through environmental legislation, not cost–benefit analysis which treats the common interest as if it were an amalgamation of private interests.

In arguing for having recourse to morality and ethics (and also aesthetics and culture) in devising environmental policy, Sagoff rejects any role at all for valuation. Environmental economists, on the other hand, would argue that the ethical enviromentalist is accommodated within their framework (Pearce and Turner 1990). Although the components of monetary value continue to be debated, in general environmental economists conceptualise these as comprising not only the satisfactions obtained from actually using an environmental resource, but also less tangible and implicitly moral satisfactions unrelated to actual or potential usage. In their taxonomies of value the pleasure an ornithologist obtains from watching corncrakes, and also from knowing that he or she will be able to watch corncrakes at any time in the future, are both use values (primary or actual use values and option values respectively). Bequest values represent a third type, for example the satisfaction derived from ensuring that corncrakes will be around for future generations to enjoy, and may be classed as use values or as non-use values. Existence values, however, are always non-use values because they relate solely to the intrinsic value of the species or habitat, irrespective of any use made of it by individuals. Corncrakes may be valued simply because they are there and ought to continue to exist. (The details of taxonomies proposed in the literature are varied; the terms used here follow Pearce and Turner (1990) and Turner et al (1992).)

This categorisation of value in terms of utility reflects the assumption that nature can be treated as if it were a set of commodities. This assumption too has its critics, particularly among conservationists. Bilsborough (1992) points out that species and habitats function not in isolation but as part of an ecosystem. Because nature's components are interdependent it is simply not sensible to divide the natural world into separate bits. For example, the UK series of SSSI forms a network of representative habitats which together secure national nature conservation goals. The value of any one site resides not just in its particular features but also in its contribution to the whole, in the role it plays in sustaining species, communities and ecosystems.

Moreover, since the market is hypothetical, the 'buyers' (the CVM survey respondents) have to be given a 'detailed description of the good(s) being valued' (Mitchell and Carson 1989, p3). CVM assumes not only that nature can be rationally partitioned but also (as with all valuations by this means) that the appropriate information is available on which to base a description of the bits in question. This assumption is also criticised by conservationists (Bowers 1990; 1993), who argue that the relevant scientific knowledge may be uncertain or lacking, and that interpretations may be contentious. 'In many cases there is, in fact, nothing that can be called *the* correct information' (Bowers 1993, p97, emphasis added).

More generally, since the publication of Kuhn's seminal work *The Structure of Scientific Revolutions* (1970) paved the way for sociological studies of scientific practice, the status of scientific knowledge has come to be more rigorously appraised. In particular, the special authority once unquestioningly enjoyed by science has been increasingly eroded, although the rhetoric of scientific certainty has tended to persist in the public sphere. Once seen as a source of infallible knowledge, scientific expertise is now more often formulated as plagued by uncertainty. (For a discussion of the nature of scientific knowledge see Murdoch and Clark (1994). The role of science in environmental issues is analysed in Yearly (1991).) Nonetheless, while environmental economists rely on science for information about the state of nature they still tend to assume that this knowledge is unproblematic.

In practice, however, the issue of information provision does plague CVM. The problem arises because the methodology does not define what information is required for the respondent to give a 'true' valuation. The question of what sort and how much information to provide is left to practitioners, and their difficulty is that valuations have been shown to vary with the quantity of information made available to respondents. For example, two groups of people were asked about their willingness to pay to protect SSSI in north east England from agricultural intensification. Those presented with detailed site descriptions, including information about rarities, were willing to pay much more than those given only the area, habitat type and SSSI grade (Mitchell et al 1988).

Practitioners of CVM conceptualise the problem of information provision as a matter of information 'bias', implying that this cause of deviation from the 'true' value is in principle correctable and will eventually be resolved with better survey design (for example, Garrod and Willis 1990; Carson 1991). Conservationists, on the other hand, see the problems associated with information provision as an irreparable defect, not only because of the difficulties of determining what the correct information is, but also because in any case respondents are likely to lack the sometimes considerable expertise needed to understand the significance of information concerning the ecological status and role of a species or habitat (Bowers 1993). In contrast, scientific criteria for ranking sites and species according to their ecological importance are explicit, even if open to interpretation (Usher 1986; Harrison 1993). In the conservationist view, not only is nature best safeguarded by strict environmental standards based on scientifically determined criteria and the judgements of experts, there is also a real danger that if monetary values are substituted for present practices the result will be more environmental degradation, while at the same time policy makers will be able to claim that wildlife has been fully accounted for.

This brief discussion of monetary valuation is by no means comprehensive, but it does reveal clear and apparently irreconcilable differences concerning what should determine which bits of nature are to be protected. These differences are not just about how environmental choices should be represented and resolved; claim and counterclaim about monetary valuation are part of an ongoing process of determining how nature will be represented and how choices concerning wildlife will be made. It is in this context that the use of CVM is debated, but although it is its *legitimacy* that is at stake, debate is grounded in claims concerning its *validity*. Economists assume that what CVM measures is unproblematic, even if its measurement may be difficult in practice. Critics reply that CVM measures little if anything, and certainly not the 'real' value of nature to society, be this moral or scientific. But neither the protestations of its opponents nor the justifications of its advocates proceed much past ideology in making such assertions.

In defending and applying CVM, environmental economists act as if people *do* make choices about nature as independent individuals choosing from a set of products on a supermarket shelf, and no more. But this is an unquestioned premise of neoclassical economic theory, not an empirically verified inference about how people actually behave. Nonetheless, neither is this assumption negated by simply *asserting* that people behave otherwise. In this sense Sagoff's ethically motivated citizens are just as much theoretical constructs as economists' satisfaction-maximising consumers. His premise is that people *do* resolve environmental choices solely via moral and ethical deliberation, but while protest bids provide evidence that people reject valuing nature in cash terms as inappropriate, many individuals, for whatever reason, are apparently amenable to stating what they would pay. (Jacobs (1994), while pursuing a somewhat different line of argument, makes similar points concerning the way arguments about valuation elide assumptions about how people behave with how they actually behave.) Nor does the assertion that willingness to pay has little meaning necessarily follow from arguments concerning the impossibility of proper description of nature or the inability of respondents to understand scientific information. For expert conservationists proper description means scientific description; their denial of a role for monetary valuation rests on the premise that the primary value of nature *does* reside in scientific understanding and (implicitly) that if lay people do not value nature primarily for this reason then they ought to. In short, in debates about valuation evidence concerning how people actually value nature is in somewhat short supply.

NATURE CONSERVATION

Recent work on the meanings of nature conservation (Harrison 1993; Burgess et al 1991) indicates that what people value in nature and how they come to value it are rather more complex than envisaged in any of the representations discussed above. Using in-depth group methodology (described in Burgess et al 1988a; 1988b) researchers explored the reactions of local conservation activists and lay people living in the Rainham area to a proposal to build a theme park and residential and commercial complex on Rainham Marshes SSSI in Essex. The committed conservationists did take for granted the ecological rationale against development, but members also acknowledged the pleasure they, and others, derive from nature 'being there'. Moreover, this group recognised a moral duty to conserve, and believed that the world of their children and grandchildren would be bereft if species and habitats continued to disappear. In contrast, the conservation case for Rainham contradicted the lay group's own experience and knowledge of nature. Nature for them was constituted in 'pleasurable encounters with the natural world – with foxes and vixens, kestrels and hedgehogs, herons and cormorants, squirrels and ladybirds' (Harrison 1993, p45), not the specialised and hidden flora and fauna found on Rainham Marshes. Perceived as an industrial wasteland of rubbish tips and rats, the Rainham Marshes local people knew did not provide access to the nature in which they took pleasure. At the same time, this group was not unambiguously in favour of development. They did not want the Marshes left as they were, but they felt intensely proprietorial and saw the developers as unsympathetic outsiders. As one member of this group put it: 'It's *our* ground, *our* council, *our* village. It's *our* right if we want it or not!' (Burgess et al 1991, p510).

The importance of place and the linking of nature with pleasurable experiences is borne out by other work which has shown the extent to which popular meanings concerning nature and open spaces are also bound up with life histories, feelings, and social and cultural values (Burgess et al 1988c; Harrison et al 1986). Nature is enjoyed as much for its beauty and wonder as for its constituent creatures, and enjoyed as much in a social context as in an individual one. The meanings that open spaces hold for individuals are bound up not only with their observations of common animals and plants but also with the sharing of experiences, with the childhood memories these place evoke, and with the social activities which take place in them.

This work clearly indicates that nature's popular significance resides largely in meanings and values other than those bestowed by scientific knowledge and understanding. And even when the scientific value of nature is accepted, nature's significance is acknowledged to rest also on moral and aesthetic concerns. Moreover, although echoes of economists' use and non-use values and of Sagoff's concerned citizens can be discerned, the message of this research is that while people certainly value nature highly, the ways in which values are constructed are complex, varied and arrived at through collective processes. Popular meanings are bound up with much more than just individual use or moral motivations, and popular values are grounded as much in real lives as in economic, ethical or scientific abstractions. The popular responses to the Rainham controversy provide strong evidence that assumptions which envisage individuals unfailingly making choices as nothing but consumers, or as nothing but citizens, or on the basis of nothing but scientific value, are far too simplistic. When confronted with environmental choices, people will draw on far richer arrays.

But while this work undermines the theoretical validity of CVM by refuting simplistic assumptions about how people value nature, it does not negate the proposition that in practice preferences concerning nature might be amenable to capture (in whole or in part) through asking people what they are willing to pay for it. What it does do is indicate that what CVM attempts to measure is considerably more complex than economists (and philosophers and conservationists) frequently presume. In practice, CVM's more reflective practitioners do acknowledge that valuation of anything other than straightforward actual use, such as recreation, remains problematic for them. It is often unclear in practice which components of their taxonomy are being quantified, nor is it understood whether people attach non-use values to specific sites or simply a general value to nature conservation (Green 1992; Green and Tunstall 1991; Garrod and Willis 1990).

The validity of CVM in practice – that is, the extent to which it might be able to provide an accurate and reliable measure, or partial measure, of what nature is worth – is simply not known. Moreover, it seems unlikely that recourse to further controlled testing to improve survey design, or to more experimental research on individual psychology to improve understanding of how people process information and make judgements, could provide adequate evidence one way or the other. Empirical research is needed, but the quantitative methods which economists call on to resolve the difficulties they encounter with CVM seem unlikely to resolve the issue of what it measures, if anything. (Practitioners of CVM encounter other difficulties besides those already mentioned. For example, results have been shown to be affected by the nature of the payment vehicle and the way in which bids are elicited.) What is needed is to gain a better understanding of what takes place at the time: how participants interpret the questions, how they construct their responses, and what knowledge, experiences and values they draw on to do so.

How popular meanings and values might mesh with a demand to assign a price to nature can still only be guessed at. But until what takes place when someone is confronted with a questionnaire inviting a willingness-to-pay response is better understood, claims that CVM provides a useful measure of nature's value to society will remain open to question, however many biases are detected or regressions performed. By the same token, until it is more clearly understood whether or not anything sensible is being measured, and if so what, claims that the results of valuation exercises represent nothing more than empty enumeration will also remain unsubstantiated.

CVM IN VALUING NATURE

Where, then, does this leave CVM as a means of valuing nature? I have argued that although there is considerable uncertainty concerning what it might measure, this does not negate the principle of monetary valuation. In so far as claims that monetary valuation should not be used in making environmental choices rest on untested hypotheses concerning how people value nature and whether or not they conceive of it as a consumer good to which market-type responses are appropriate, the question of its legitimacy remains unresolved. While the theoretical assumptions of the neoclassical approach are clearly inadequate, it nonetheless remains possible that in practice CVM's monetary valuations could provide a useful measure, at least in part, of the worth of the natural world to society.

Monetary valuation of environmental resources in general, and nature in particular, is justified by environmental economists on the grounds that such non-commoditised resources would otherwise be undervalued in decision making, and undervaluation results in a lower level of environmental protection than is actually desired by society. Incorporating the demand for environmental resources into CBA, as measured by aggregating individual willingness to pay for them (or by some other method), redresses this failure. Reliable valuations are presumed to be available.

Reliable valuations, reliable in the sense that they satisfactorily reflect what individuals choose, would not of themselves legitimise the environmental economist's case. CBA presumes the superiority of economic efficiency. The case against economic efficiency *per se* has been made by both economists (see Bromley 1990) and philosophers (not least Sagoff 1988a) and encompasses the various internal difficulties, both theoretical and practical, associated with operationalising the Pareto criterion. (These difficulties are well recognised within the economics profession. Bromley (1990) argues that economic efficiency, judged by the Pareto test, 'fails the test of consistency and coherence within economic theory' (p104).) There is also the argument, to quote Bromley, 'that *it is a value judgement* for the economist to claim that economic efficiency *ought to be* the decision-rule for collective action' (p97). Since, it is argued, this value judgement is not supported by society as a whole, recourse to economic efficiency is not legitimate in public policy making.

The environmental economist's case, however, also has a more pragmatic basis. Recourse to environmental cost–benefit analysis is argued to ensure better protection of the environment not just because economic efficiency ought to count but because it *is* what counts in the real world of decision making. But there is no reason why this should necessarily ensure that nature is *better* protected than it otherwise might be. As O'Neill (1993) points out, this will only happen if individual preferences already happen to be the 'right' ones. The monetary valuation (or shadow price) for the bit of nature in question has to exceed that for the good which requires its destruction. 'An environmentally friendly outcome depends on the shadow price picking up a predominance of strong preferences for environmental goods. If people prefer marinas to mud flats, Disneyland to wetlands, and roads to woodland, then no amount of shadow pricing will deliver environmentally friendly results' (p77). This line of thinking forms part of a complex web of reasoning (to which I have not attempted to do justice here) deployed by O'Neill to reject the 'want-regarding principles', which lie at the heart of neoclassical analysis, as incapable of protecting the environment. He defends instead an 'ideal-regarding' position which gives central place to the sciences and the arts in decision making about the environment.

These arguments deny the legitimacy of monetary valuation in absolute terms, giving no quarter to economic analysis even were monetary valuations satisfactory. The problem for the pragmatist is that economic analysis *does* count in environmental choices (Corkindale 1993; Department of the Environment 1991) and valuation often accords a higher cash value to the natural world than would otherwise be the case, even if it does not always ensure its protection. Moreover, one merit of CVM would seem to be that it actually involves the public in environmental policy in a direct way, unlike elite political processes or professional scientific evaluation. If monetary valuation is accepted on pragmatic grounds, why not use CVM for valuing nature, provided it does satisfactorily capture what people want?

The legitimacy of using CVM rests not only in greater certainty concerning what it measures, but also in its satisfactory translation of popular preferences. Estimates derived via CVM as currently practised appear to deny work on popular values which indicates that nature is *highly* valued, because its practices refute *a priori* the credibility of high figures. Guidelines for CVM enshrine conservative estimates. They recommend that survey questions use options known to 'underestimate' willingness-to-pay figures, and that data analysis should exclude extreme responses because they increase estimated values 'wildly and implausibly' (NOAA 1993, p4612).

Secondly, and more fundamentally, the legitimacy of CVM is called into question because of the way in which the survey and the issues at stake are presented to respondents (see, for example, Willis et al 1993). Valuation is usually presented as nothing more than a matter of paying so much for the specified bit of nature via taxes, donations, permits or some other vehicle. The issues are often framed in terms of conserving or protecting a site or a species, rather than in terms of policy alternatives such as the choice between keeping a woodland and building a road through it. Most importantly, perhaps, the way in which responses will be used tends not to be made explicit. CVM surveys are presented as surveys, not as instruments which may have implications for policy in the real world. As long as the terms of participation are not made explicit, and respondents are not given the information they would need to consent to the process, CVM's whiff of democracy must remain more apparent than real. In these circumstances, any prices put on nature would seem to be suspect.

CONCLUSION

It might seem that I have argued the case against CVM in a peculiarly round-about way. It is more usual to condemn monetary valuation *per se*, in which case CVM goes too, or to reject as unsound the assumptions of neoclassical theory which underlie CVM. The outlines of some of these arguments have been presented in the course of this discussion. But as a pragmatist, and as someone who would defend participatory processes, I find it difficult to repudiate CVM so absolutely, and it is this which compelled me to a closer examination of what it measures and how it is practised. Clearly, it is not the case that any valuation is better than no valuation. The proviso is that non-market valuations are meaningful, otherwise the whole edifice falls. And in practice CVM seems to fall, as much as anything, on the deceits of its operation. Not only is it necessary to understand better what CVM surveys measure in order to decide whether such valuations are useful, it will only be legitimate to use the method if respondents are properly informed about the process in which they are participating and about any potential policy implications. And if they are fully informed, will they then respond by putting a price on nature?

REFERENCES

Bateman, I. (1993) 'Consistency between Contingent Valuation Estimates: A Comparison of two UK National Parks', ESRC Countryside Change Initiative Working Paper 40, Department of Agricultural Economics and Food Marketing, University of Newcastle upon Tyne, Newcastle.

Bilsborough, S. (1992) 'The Oven-Ready Golden Eagle: Arguments Against Valuation', *ECOS*, **13**, 1, 46–50.

Bowers, J. (1990) *Economics of the Environment: The Conservationists' Response to the Pearce Report*, British Association of Nature Conservationists, Shropshire.

Bowers, J. (1993) 'A Conspectus on Valuing the Environment', *Journal of Environmental Planning and Management*, **36**, 1, 91–100.

Bromley, D. (1990) 'The Ideology of Efficiency: Searching for a Theory of Policy Analysis', *Journal of Environmental Economics and Management*, **19**, 86–107.

Burgess, J., Limb, M. and Harrison, C. M. (1988a) 'Exploring Environmental Values through the Medium of Small Groups: 1 Theory and Practice', *Environment and Planning A*, **20**, 309–26.

Burgess, J., Limb, M. and Harrison, C. M. (1988b) 'Exploring Environmental Values through the Medium of Small Groups: 2. Illustrations of a Group at Work', *Environment and Planning A*, **20**, 457–76.

Burgess, J., Harrison, C. M. and Limb, M. (1988c) 'People, Parks and the Urban Green: A Study of Popular Meanings and Values for Open Spaces in the City', *Urban Studies*, **25**, 455–73.

Burgess, J., Harrison, C. and Maiteny, P. (1991) 'Contested Meanings: the Consumption of News about Nature Conservation', *Media, Culture and Society*, **13**, 499–519.

Carson, R. T. (1991) 'Constructed Markets' in J. B. Braden and C. D. Kolstad (eds) *Measuring the Demand for Environmental Quality*, Elsevier, Amsterdam.

Cole, K., Cameron, J. and Edwards, C. (1991) *Why Economists Disagree: The Political Economy of Economics*, Longman, London.

Corkindale, J. (1993) 'Recent Developments in Environmental Appraisal', *Journal of Environmental Planning and Management*, **36**, 1, 15–22.

Department of the Environment (1991) *Policy Appraisal and the Environment: A Guide for Government Departments*, HMSO, London.

Garrod, G. and Willis, K. (1990) 'Contingent Valuation Techniques: A Review of their Unbiasedness, Efficiency and Consistency', ESRC Countryside Change Initiative Working Paper, Department of Agricultural Economics and Food Marketing, University of Newcastle upon Tyne, Newcastle.

Green, C. (1992) 'The Economic Issues Raised by Valuing Environmental Goods', in A. Coker and C. Richards (eds) *Valuing the Environment: Economic Approaches to Environmental Valuation*, pp28–61, Belhaven Press, London.

Green, C. H. and Tunstall, S. M. (1991) 'Is the Economic Evaluation of Environmental Resources Possible?', *Journal of Environmental Management*, **33**, 123–41.

Harrison, C. (1993) 'Nature Conservation, Science and Popular Values', in F. B. Goldsmith and A. Warren (eds) *Conservation in Progress*, pp35–49, John Wiley, Chichester.

Harrison, C., Burgess, J. and Limb, M. (1986) 'Popular Values for the Countryside', report prepared for the Countryside Commission, Department of Geography, University College London, London.

Jacobs, M. (1994) 'The Limits of Neoclassicism: Towards an Institutional Environmental Economics', in M. Redclift and T. Benton (eds) *Social Theory and the Global Environment*, Routledge, London.

Kuhn, T. (1970) *The Structure of Scientific Revolutions*, University of Chicago Press, Chicago.

Mitchell, L. A. R., Willis, K. G. and Benson, J. F. (1988) 'A Review and Empirical Tests of Bias in Contingent Valuations of Wildlife Resources', Working Paper Series No. 5, Department of Town and Country Planning, University of Newcastle upon Tyne, Newcastle.

Mitchell, R. C. and Carson, R. T. (1989) *Using Surveys to Value Public Goods: the Contingent Valuation Method*, Resources for the Future, Washington, DC.

Murdoch, J. and Clark, J. (1994) 'Sustainable Knowledge', *Geoforum*, **25**, 115–32.

NOAA (National Oceanic and Atmospheric Administration) (1993) 'Natural Resource Damage Assessments Under the Oil Pollution Act of 1990', *Federal Register*, **58**, 10 (15 January), 4601–14.

O'Neill, J. (1993) *Ecology, Policy and Politics: Human Well-Being and the Natural World*, Routledge, London.

Pearce, D. W. (ed) (1991) *Blueprint 2: Greening the Global Economy*, Earthscan, London.

Pearce, D. W. and Turner, R. K. (1990) *Economics of Natural Resources and the Environment*, Harvester Wheatsheaf, London.

Pearce, D., Markandya, A. and Barbier, E. B. (1989) *Blueprint for a Green Economy*, Earthscan, London.

Sagoff, M. (1988a) *The Economy of the Earth*, Cambridge University Press, Cambridge.

Sagoff, M. (1988b) 'Some Problems with Environmental Economics', *Environmental Ethics*, **10**, 56–74.

Sagoff, M. (1993) 'Environmental Economics: An Epitaph', *Resources for the Future*, **111**, 2–7.

Tunstall, S. M. and Coker, A. (1992) 'Survey-based Valuation Methods' in A. Coker and C. Richards (eds) *Valuing the Environment: Economic Approaches to Environmental Valuation*, pp104–26, Belhaven Press, London.

Turner, R. K., Bateman, I. and Brooke, J. S. (1992) 'Valuing the Benefits of Coastal Defence: a Case Study of the Aldeburgh Sea-defence Scheme', in A. Coker and C. Richards (eds) *Valuing the Environment: Economic Approaches to Environmental Valuation*, pp77–100, Belhaven Press, London.

Usher, M. B. (1986) 'Wildlife Conservation Evaluation: Attributes, Criteria and Values', in M. B. Usher (ed) *Wildlife Conservation Evaluation*, pp4–44, Chapman and Hall, London.

Wilks, L. C. (1990) 'A Survey of the Contingent Valuation Method', Resource Assessment Commission Paper No. 2, Australian Government Publishing Service, Canberra.

Willis, K., Garrod, G., Saunders, C. and Whitby, M. (1993) 'Assessing Methodologies to Value the Benefits of Environmentally Sensitive Areas', ESRC Countryside Change Initiative Working Paper, Department of Agricultural Economics and Food Marketing, University of Newcastle upon Tyne, Newcastle.

Yearly, S. (1991) *The Green Case: A Sociology of Environmental Issues, Arguments and Politics*, HarperCollinsAcademic, London.

7 Environmental Values as Discursive Resources

KATE BURNINGHAM
Centre for Environmental Strategy, University of Surrey

INTRODUCTION

Much social research into environmental issues concentrates on the area of environmental values and attitudes, asking questions such as how they are distributed in the population, what influences them and what effects they have on practice. In this chapter I apply the recommendations of discourse analysis (Gilbert and Mulkay 1984; Potter and Wetherell 1987; Edwards and Potter 1992) to the area of environmental values. From this perspective, rather than being considered as something which inheres in individuals, environmental values may be more productively viewed as a resource in language. This position is illustrated by some data from interviews with participants in disputes about proposed road schemes in which interviewees use environmental values strategically to justify their claims and complaints.

Existing psychological work tends to focus on environmental attitudes rather than values. There is a body of such work which aims to identify the determinants of environmental attitudes and their relation to pro-environmental behaviour such as recycling. This research examines the influences of a wide range of factors such as gender, personality, level of education and environmental knowledge on environmental attitudes. The conclusions are often contradictory, and there is no consensus about which factors are related to the espousal of environmental attitudes, or whether those with such attitudes are more likely to engage in action to protect the environment.

The most prominent work on environmental values was carried out in the 1970s and 1980s by the American sociologists Dunlap et al (eg Dunlap and Van Liere 1978; 1984; Dunlap et al 1981). They suggested that those who espouse environmental attitudes have different value orientations to those who do not, and argue that a shift is occurring in the values prevalent in society from what they term the human exceptionalist paradigm (HEP), which stresses growth and mastery over nature, to the new environmental paradigm (NEP), which values nature highly and accepts limits to growth. Following on from this work others have attempted to specify the value orientations which give rise to environmen-

Values and the Environment: A Social Science Perspective, Edited by Yvonne Guerrier, Nicholas Alexander, Jonathan Chase and Martin O'Brien. © 1995 John Wiley & Sons Ltd.

tal attitudes and perhaps behaviour. For example, it has been suggested that those with altruistic value orientations are likely to engage in pro-environmental behaviour (eg Hopper and Nielsen 1991; Oskamp et al 1991). Stern et al (1993) have recently added to this theory claiming that egoistic value orientations and value orientations towards the welfare of non-human species or towards the biosphere itself are also implicated in environmental attitudes and behaviour.

Research in this tradition generally seeks to uncover whether individuals hold environmental attitudes or values by asking them questions in surveys or interviews. A variety of methodological approaches are used, but all share the assumption that the answer given, whether in the form of a mark on a Likert scale or a detailed account, provides some insight into the individuals' mental state – their attitude or value orientation. However, there is a growing tradition of work within linguistics, social psychology and sociology which identifies problems with this assumption. A variety of approaches have been developed which take issue with the notion that language merely provides a detached commentary on reality. In this chapter I consider how the approach known as discourse analysis can be applied to reconceptualise the study of environmental values and attitudes.

DISCOURSE ANALYSIS

Although the term 'discourse analysis' is used to refer to a variety of analytic approaches in sociology, sociolinguistics and social theory, in this paper it will be used only to refer to the approach exemplified by Gilbert and Mulkay (1984) and Potter and Wetherell (1987). This approach has its origins in the philosophical work of Wittgenstein and Austin, and is influenced by the sociological recommendations of ethnomethodology and conversation analysis. From this perspective language is not regarded as 'standing for' states of affairs as they 'really are', but as an active medium through which this reality is actually constituted. The insights of discourse analysis have radical implications for the methodologies traditionally used to gauge people's environmental attitudes or values. The two issues that have particular relevance are the realisation that there may be immense variability within and between accounts, and the recognition that accounts are constructed by speakers to attend to specific interactional functions.

In their study of scientists' actions and beliefs Gilbert and Mulkay noted that within the accounts they gathered there was some variability in the way that the same events were described. This was apparent not only in accounts from different scientists, or from the same scientist in different contexts (eg in journal papers, interviews and informal conversation), but also within the same interview. That is, individual scientists regularly characterised events and their beliefs about them in quite contradictory ways during the course of an interview. Such variability is a pervasive feature of research which relies on accounts and clearly poses a problem for forms of analysis which seek to provide a definitive version of action or belief.

Potter and Wetherell, working in social psychology, extended Gilbert and Mulkay's findings to the study of attitudes. In their interviews with white middle-class New Zealanders on generally controversial issues in New Zealand society, Potter and Wetherell (1987) found that within the same interview respondents made some statements which seemed to indicate they thought that immigrant Polynesians should leave the country, and others which suggested they were happy for them to stay in New Zealand. These sorts of observation throw doubt on the assumption that when people report on their attitudes, values or views, they are performing a neutral act of describing an inner mental state. Instead they suggest that people are doing things in and with their spoken and written statements. For instance, they might be making complaints, justifying their position, blaming others, seeking sympathy or approval. All of these are actions. Discourse analysis suggests that rather than attempting to use accounts to interpret and describe some reality which is deemed to exist independently of them, it is more fruitful to analyse the accounts themselves, to explore how specific versions are formulated and to what ends.

Following the recommendations of this approach a study of environmental values puts aside the question of whether individuals 'really' hold environmental values or not, and concentrates instead on how such values are used within specific accounts. In the remainder of this chapter I will outline what a discourse analytic approach to the study of environmental values might look like, by presenting data collected during two research projects into the impacts of road schemes. It should be emphasised that owing to restrictions of space what follows are preliminary analytic remarks designed to illustrate the approach, rather than a thorough analysis of the data.

DATA

The data discussed in this chapter comes from two projects into the social impacts of trunk road schemes. The first project (1990–91) was funded by Acer Consultants Ltd, a group of consultant engineers who had been contracted by the Department of Transport to assess and design alternative routes for a proposed road 'improvement' scheme in the south of England. Acer subcontracted the social impact component of the assessment to sociologists at the University of Surrey. The second project was funded by the Economic and Social Research Council (ESRC), and aimed to provide a post-impact study of an area where a trunk road had recently been completed. The road studied was another stretch of the trunk road examined in the Acer study and had been open to traffic for five years.

As part of the Acer project, interviews were carried out with local informants. These were people who were expected to be able to provide an overview of local concerns such as local councillors, clergy and headteachers as well as leaders of residents' associations, action and conservation groups and others who could be expected to be adversely affected by one of the schemes under consideration. In addition letters submitted during previous consultation exercises, press reports and pamphlets and statements produced by action groups were collected. Excerpts from the interviews and action group publications provide some of the data to be analysed. The ESRC project also involved interviews with local people who were involved in the debate about the road before it was built or had been active in campaigning about aspects of it since it opened to traffic. Extracts from some of these interviews will also be discussed.

ENVIRONMENTAL VALUES VS NIMBY COMPLAINTS

The issue of the siting of new roads, and whether they should be built at all, provides a clear and accessible arena for studying the production and use of environmental values. There is some debate about whether those who object to the prospect of developments in their locality do so on the basis of broad environmental values or narrow NIMBY (not in my back yard) concerns. For instance, Kemp (1990) argues that those objecting to the local dumping of radioactive waste are motivated by a range of environmental concerns and should not simply be characterised as NIMBY. This debate is also evident in media accounts. For instance, one recent newspaper article (Nicholson-Lord, *The Independent*, 20 June 1993) cites a representative of Alarm UK (a group which represents 150 local anti-road groups) who claims that 'Roads and traffic have become the focus of wider environmental concern', while another recent report documenting local protest about a proposed road scheme in Dorset argues that 'there is connection between green values and the self interest of those who invoke them'(Wright, *The Guardian*, 12 June 1993). Following the recommendations of discourse analysis I take an alternative approach. Rather than using respondents' statements to try to discern whether their complaints are motivated by environmental or by NIMBY values, I examine how they use such 'values' in their accounts and to what ends.

When new developments such as road schemes are proposed the issue often becomes the site for local debate and conflict. There is rarely consensus about how the project will affect the locality, what the significant impacts might be or how they might best be ameliorated. Differences of opinion exist not only between the planners and consultants responsible for designing and assessing the scheme and the local population, but also between different individuals, groups and organisations and even between 'experts'. Participants in such disputes engage in a variety of strategies to present their position as more credible, robust and convincing than that of others. The practices which individuals and groups deploy to warrant the authenticity of their own account, or to suggest that another's version is untrustworthy, are central to the dispute. The issue of whether an account appears to be factual, is robust, or insidiously undermines alternative accounts is of prime importance to participants – as Edwards and

Potter (1992) put it, it is a 'live issue' for them. It is not simply an issue of academic interest but is something of which participants are aware and take account. They recognise that the way in which they construe the costs and benefits of alternative schemes may have practical implications for the outcome of the dispute. In the remainder of this chapter I will outline how the imputation of values may be considered as one such discursive strategy, drawing attention to the way in which participants present their own position as informed by environmental values while those of others are denigrated as motivated by NIMBY concerns.

Analysis of a variety of different sorts of accounts has illustrated the attention which speakers pay to presenting their version as objective, unbiased and considered in contrast to the ill-informed, biased or misguided views of others (see Gilbert and Mulkay 1984; Edwards and Potter 1992). For example, in their interviews with scientists Gilbert and Mulkay identified a phenomenon which they called 'accounting for error'. They called an example of 'accounting for error' any account in which the speaker identifies the views of one or more scientists as mistaken and provides some kind of account which enables us to understand why these scientists adopted an incorrect theory. In these accounts speakers linked their own position firmly to empirical facts such as experimental results, while the positions of those they identified as being in error were explained in highly contingent terms; for example as the result of stubbornness, stupidity, bias or prejudice. Gilbert and Mulkay write that these accounts have an asymmetrical structure: speakers employ an empiricist repertoire to account for their own position, and a contingent one to account for the positions of others. In the empiricist repertoire scientists characterise their actions and beliefs as following unproblematically from the empirical facts, while in the contingent repertoire actions and beliefs are accounted for in terms of factors such as personal characteristics, social ties or group membership.

In the data collected during my research into the impacts of road schemes a similar pattern is apparent. Speakers attribute the 'wrong' positions of others to the result of 'putting their head in the sand', 'not listening to facts', 'not understanding' and personal interest, while their own position is characterised as 'facing up to facts', objective, based on common sense and 'rational appraisals' of expert advice and information. Clearly ordinary people rely on empiricist repertoires to bolster their case, and contingent repertoires to undermine the position of others, just as much as scientists do. However, in these accounts there is an additional strategy which exists alongside and complements attempts to portray their own position as impartial and others as biased. Speakers not only attempt to render their position credible and 'right' by asserting its objective and unbiased nature, but also by stressing that it is informed by environmental values. Similarly, their opponents' views are not only denigrated as ill informed but are depicted as unenvironmental; their proposals are characterised as both damaging to the environment and motivated by self-centred NIMBY concerns.

CONSTRUCTING OWN POSITION AS ENVIRONMENTAL

In the Acer study the choice of route for the road was essentially between an on-line improvement passing through a residential area, or a bypass across downland, part of which was designated an area of outstanding natural beauty (AONB). Whichever route individuals or groups supported, they claimed to do so on the grounds that this was the best environmental option. Participants claimed not only that the route they opposed would have adverse effects on the environment, but also where they were actively supporting the construction of another route, they claimed that this would actually have *beneficial* effects on the environment. For example, in pamphlets produced by groups opposing the on-line route in favour of a bypass, it was claimed that the route they opposed would 'have an unacceptable impact on the environment', whereas a downland bypass would 'enable others to share with us the splendour of the views which the Downs provide'.

Those who hold the opposite position, opposing a downland route in favour of the on-line alternative, used exactly the same strategy. One interviewee claimed that a bypass would 'ruin the Downland environment', while the on-line route would improve the environment in the residential area by lessening the pollution caused at present by traffic congestion:

> Yes, the environment in [*town*] would change, but I cannot believe that it could be worse. Because at the moment with the congestion . . . exhaust fumes are beyond the maximum that one should have . . . so the pollution of the air is infinitely greater now by the non movement of the traffic than it would be if a road were put in and traffic were able to move freely.

This is not just a tactic employed by local people but one which permeates the whole debate. The Department of Transport and its consultants are also at pains to prove that their preferred route is the best environmental option, stating clearly in a statement explaining their proposals that the scheme was chosen because 'the impact on the landscape would be less [than for other routes,]' and 'the route would have no significant impact on important flora and fauna'.

In addition, this strategy is not confined to the local discussion of this particular scheme, but is employed in national debate about transport policy. In 1992 the Transport Secretary, John MacGregor, characterised new roads as being 'good for the environment' (*The Guardian*, 1 June 1992), and more recently a spokeswoman for the Department of Transport claimed that when new roads are planned the Department aims 'always to minimise adverse aspects on the environment' (*The Independent*, 25 June 1993).

Such assertions are at odds with the claims of environmental activists and organisations that the road-building programme has adverse environmental impacts from the local to the global level.

DENIGRATING OPPONENTS' POSITION

Just as participants in such disputes are keen to portray their position as motivated by wide environmental values, so they denigrate the position of those with different views as motivated by narrow selfish values. The following extract is taken from an interview carried out with members of a residents' association during the Acer project. The association favoured an on-line route, or failing that a bypass south of the AONB rather than a bypass further north.

> He [local MP] said and I quote . . . 'a bypass route . . . I think that would have serious disadvantages'. Now you know why don't you? It goes through his Golf Club. He's president of the golf club, so it mustn't go through there.

In this extract the speaker suggests that the local MP is only opposed to the building of a southern bypass because that route would affect land belonging to the golf club of which he is president. His position is effectively undermined by the suggestion that it is based on selfish concerns. Later in the same interview another member of the group stresses that their own position is not based on such a limited concern to protect the area in which they live, but on wider and more objective concerns:

> [the association] is . . . a number of people who are concerned, who've got together and thought we must put the other side of the case, we must make sure it's presented. And not from the view of [village] but from the point of view of just ordinary common sense logic and economics.

Thus a clear contrast is presented in the interview between the selfish basis of the complaints of others, and the impartial basis of the group's own complaints. Research in a variety of different settings (eg Smith 1978, in an account of a young woman's psychiatric problems; Atkinson 1984, in political speeches) has illustrated the way in which contrasts are often employed in accounts in this way to bolster a particular version of events or behaviour, and undermine others.

In the next excerpt the speaker also constructs a clear contrast between his own objective views and the narrow self-interested complaints of others. The extract is taken from an interview conducted during the ESRC project in which the leader of a residents' association is reflecting about the public inquiry held before the road was built.

> Well, there was a deal of controversy about it [the public inquiry] because the various factions representing the route alternatives were all quite articulate, and I think made their cases very well. The Inspector had a tough job really arbitrating between the various objections, because they were six of one and half a dozen of the other, and I think it would be fair to say that the underlying objective of the people that spoke was to keep the thing away from their personal patch. Understandably, but it's the old story. We had the advantage in that most of the people who represented the

Association . . . were not directly affected by the road, and therefore could
remain reasonably objective about the route alternatives.

He credits other objectors at the public inquiry with having given the inspec-
tor a 'tough job' as they were all 'quite articulate', however their arguments are
subsequently undermined by the suggestion that these people were motivated
simply to 'keep the thing away from their personal patch'. The speaker sets up
the majority of other objectors at the inquiry as being motivated by self-interest,
and then contrasts them with the position of the association to which he
belongs, which he claims could be 'reasonably objective' because its members
were not directly affected by the road.

AVERTING NIMBY CHARACTERISATIONS

In both interviews the speakers appeared to recognise that their own position
may be as susceptible to charges of narrow self-interest as those which they
criticise. They acknowledge this possibility and work to avert this categorisation. In the first interview the speaker states clearly that their position is not
'from the point of view of [village]' and in the second the speaker claims that
their position could not be denigrated as self-interested as none of the members
of the association expected to be directly affected by the road. This deliberate
work by interviewees to avert the inference that their concerns are motivated by
self-interest is clear in other interviews. For example, the following extract is
taken from an interview conducted during the Acer research, in which a man
who lived and ran a school close to one of the proposed routes for the road was
outlining the basis for his opposition:

> Just because we happen to live in this area we shouldn't say, 'oh no, well
> nobody else could come here, you know, and sort of enjoy it'. But the point
> I think is, that what we feel is that it has been open country always, and
> once it's changed it's changed for ever. Gone. More town. And very soon
> there won't be any sort of open space along the South Coast at all, you
> know, it'll all be bricked over.

In this extract he acknowledges that protecting the area in which one lives
may be considered selfish and states that complaints based on this motivation
'shouldn't' be made. He goes on to outline the broad base for his complaint: his
concern that the countryside is being eroded by the construction of new roads
and consequent development around them. He constructs his concern as being
about the whole South Coast rather than simply about the area in which he
lives, or the particular road being proposed.

CONCLUSION

Clearly, local residents who object to plans for new road schemes are in a deli-
cate position. They are bound to be concerned to protect the environment in

which they live, and their knowledge about this constitutes one of the only spheres of expertise available to them (see Kemp 1990; Burningham 1993). In addition, the structure of the participation process precludes discussion of wide environmental and transport policy issues at the public inquiry, and rules that the only legitimate discussion is of local impacts of the proposed road scheme. Thus for a variety of reasons participants will want to express their concern about their 'patch', but they are aware in doing so that their complaints may be undermined as NIMBY. As a consequence we see in their accounts deliberate attempts to construct their position as founded on objective criteria and informed by environmental values, acknowledgements that self-interest provides an inadequate basis for complaint, and attempts to distinguish their position from others which they claim may rightly be characterised as NIMBY.

ACKNOWLEDGEMENTS

This chapter is based in part on research funded by Acer Consultants Ltd, and in part on research funded by the ESRC under award no R000233246. I would like to thank Nigel Gilbert for helpful comments on a previous draft.

REFERENCES

Atkinson, M. (1984) *Our Masters' Voices,* Methuen, London.

Burningham, K. (in press) 'Us and Them: the Construction and Maintenance of Divisions in a Planning Dispute', in C. Samson and N. South, *The Social Construction of Social Policy: Methodologies, Racism, Citizenship and the Environment,* Macmillan, London.

Dunlap, R. and Van Liere, K. (1978) 'The New Environmental Paradigm: a Proposed Measuring Instrument and Preliminary Results', *Journal of Environmental Education,* **9**, 10–19.

Dunlap, R. and Van Liere, K. (1984) 'Commitment to the Dominant Social Paradigm and Concern for Environmental Quality: an Empirical Examination', *Social Science Quarterly,* **65**, 1013–28.

Dunlap, R., Grieneeks, J. K. and Rokeach, M. (1981) 'Human Values and Pro-environmental Behavior', in W. Conn (ed) *Energy and Material Resources: Attitudes, Values and Public Policy,* Boulder, Westview Press.

Edwards, D. and Potter, J. (1992) *Discursive Psychology,* Sage, London.

Gilbert, G. N. and Mulkay, M. (1984) *Opening Pandora's Box: a Sociological Analysis of Scientists' Discourse,* Cambridge University Press, Cambridge.

Hopper, J. and Nielsen, J. (1991) 'Recycling as Altruistic Behaviour: Normative and Behavioral Strategies to Expand Participation in a Community Recycling Program', *Environment and Behavior,* **25**, 195–220.

Kemp, R. (1990) 'Why Not in my Backyard? A Radical Interpretation of Public Opposition to the Deep Disposal of Radioactive Waste in the United Kingdom', *Environment and Planning A,* **22**, 1239–58.

Oskamp, S., Harrington, M. J., Edwards, T. C., Sherwood, D. L., Okuda, S. M. and
    Swanson, D. C. (1991) 'Factors Influencing Household Recycling Behavior',
    Environment and Behaviour, **23**, 494–519.
Potter, J. and Wetherell, M. (1987) *Discourse and Social Psychology: Beyond Attitudes
    and Behavior,* Sage, London.
Smith, D. (1978) 'K is Mentally Ill: the Anatomy of a Factual Account' *Sociology,* **12**,
    23–53.
Stern, P., Dietz, T. and Kalof, L. (1993) 'Value Orientations, Gender and
    Environmental Concern', *Environment and Behavior,* **25**, 322–48.

Section III

POLICY DILEMMAS

Introduction

NICHOLAS ALEXANDER
Faculty of Business and Management, University of Ulster

In the four chapters in this section, the authors address some of the policy dilemmas which face environmental planners and legislators. The chapters provide an insight into the complexity of environmental issues and the sectional interests which inform and shape the environmental debate.

The simplistic interpretations with which public support is often courted must be understood, both in terms of the issues as they are evaluated and recognised by diverse groups, and how such evaluations are interpreted in the light of the sectional interests of such groups. Sectional interests, along with vested interests, are mirrors which distort even the most apparently altruistic of intentions. The common good is as elusive in environmental policy decisions as in other policy areas with which the body politic is concerned. The road to an environmental hell may well be paved by individual misuse of the environment, but equally, as the issues discussed in this section illustrate, it may also be paved by good intentions.

The debate on general environmental concerns, and specific environmental issues, is tempered by a philosophical understanding of the debate, but it is also qualified and quantified on the basis of measurement systems which may complement, and may contradict, fundamental assumptions. While Section II of this book considered epistomological issues, the chapters in this section also provide an insight in this respect.

The dilemma addressed by Pearson in Chapter 8 is at the international level and concerns the domestic problems of individual states together with the problems shared by groups of states. This dilemma is discussed in terms of the issues associated with energy consumption and use, and its consequential effect on economic growth. Developing countries and industrialised countries, while sharing a planet, do not face the same economic issues in respect of energy use. Developing and industrialised countries, therefore, do not share the same priorities.

While Pearson acknowledges that generalisations about shared interests should be treated carefully, it is nonetheless possible to agree that the concerns of developing countries provide them with a dilemma and, to some extent, a common set of problems which inhibit the acceptance of an agenda established by industrialised countries. Indeed, fundamental to the difference in priorities is

Values and the Environment: A Social Science Perspective, Edited by Yvonne Guerrier, Nicholas Alexander,
Jonathan Chase and Martin O'Brien. © 1995 John Wiley & Sons Ltd.

the identification of fundamental environmental problems. While industrialised countries may, through an evaluation of their own experience, consider energy as a, if not the, fundamental environmental problem, this may be far from the case in those developing countries where other, sometimes severe, problems within the environment may be deemed to be either of greater significance or simply of importance where the energy issue is not.

Pearson gives a useful matrix for exploring the priorities and trade-offs appropriate to the issue addressed in his chapter and also provides a framework within which to consider the other chapters in this section. The matrix classifies environmental issues on the basis of the spatial and temporal context.

McKenzie Hedger in Chapter 9 continues the energy theme. In this chapter the issues of wind farms in the UK, and particularly Wales, are considered. Where Pearson's chapter addressed the concerns of developing countries in the light of the global environmental agenda, this chapter considers the problem of introducing into communities an alternative energy-generating system which has been perceived as beneficial within the wider socioeconomic framework. As Pearson draws out the fundamental and different set of priorities which will affect the interpretation of environmental problems on an international level, McKenzie Hedger's chapter explores the different priorities and arguments of interested parties within the national context and at the local level.

The cases of the supporters and opponents of wind farm projects are represented and explored. Like Pearson's chapter, Chapter 9 provides a valuable insight into the dilemmas which face planners and the different sets of priorities which sectional interest groups have and in consequence of which they seek to set the environmental agenda. McKenzie Hedger's conclusions raise important issues as to the means by which such arguments may be reconciled and by whom reconciliation may be achieved.

Doupé in Chapter 10 considers the judiciary's response to environmental problems through the interpretation of the laws of England. Like McKenzie Hedger, he focuses on a particular issue in order to explore the complexity of environmental problems. In this chapter, the issue is water pollution and the rights of industrial companies, and efforts by the 'guardians' of the environment to control pollution levels.

Doupé's chapter raises the issue of the effect of changing legislation and its effects on industrial operations which have previously been carried out without contravention of the law. It also raises the issue of legislation and remedy. In the case considered, although extensive legislation exists at the UK and European level, the plaintiffs in the case found recourse to the common law to be their best approach. Given the nature of common law, and hence the role of the judiciary, the judiciary's response to environmental issues is emphasised.

Doupé contrasts the attitudes of the nineteenth-century judiciary with those of the twentieth century in interpreting the 'non-natural' use of the environment. It was on this basis that an initial decision, based on common law, could appear contrary to European legislation. The values of the judiciary within the national context and environmentally concerned legislators at the supranational level are neatly counterpoised.

Bonnes and Bonaiuto in Chapter 11 consider the evaluation of the urban environment, both natural and built, from the perspective of the expert and layperson. These terms in themselves raise important questions of environmental assessment. Echoing Doupé's discussion of the 'natural' use of the environment, this chapter considers the natural in the context of the 'civis' and juxtaposed with the built environment. Technical assessment of the environment, which is associated with the expert, is 'objective' and based on measurement systems, whereas 'subjective' assessment, while used by experts in their evaluation of the environment, is descriptive and thereby judgemental and is essentially the response of the layperson. However, as Bonnes and Bonaiuto recognise, the distinction is far from clear.

This chapter recognises, and in so doing returns to a theme in the other chapters in this section, the problems of individuals' and groups' world view of environmental problems, and the apparent inconsistency with such expressions of opinion at the local level. This, as they note, is a product of a lack of common understanding of terms used in the environmental debate but also the multidimensional nature of the discussion which is the product of various evaluative perspectives: recreational, aesthetic, ecological and functional.

Environmental threats are both measurable in the scientific sense and observable in the perceptual sense. However, as Bonnes and Bonaiuto note, while scientific and perceptual measures may have some relationship, perceptions of pollution are often expressed in scientific terms, while an indication of environmental threats is, on occasion, better realised through attitudes to the perceived threats and the relationship of natural and non-natural elements within the landscape; an issue which may be profitably considered in the light of Doupé's consideration of the English judiciary's attitude to pollution and the tensions inherent in the relationship between objective and subjective measurement.

Bonnes and Bonaiuto's chapter focuses on the quartiere Aurelio in Rome. The material discussed is drawn from laypersons' responses to questionnaires which considered 'spatio-social density', 'socio-spatial security', and 'socio-functional facilities', along with expert evaluations of the built and natural environments. Further consideration of laypersons' evaluation of 'spatiosocial density' enabled direct comparison with the evaluations of the expert group. The results emphasise the problems inherent in evaluations of environmental values of experts and laypersons. This bears on the issues raised in McKenzie Hedger's chapter. Observational evaluation by the layperson, which might be considered subjective, will, as the debate gathers momentum, begin to eclipse the measurable conclusions of the experts, which might be considered objective.

The competing interests and evaluation criteria of the different groups considered in these chapters emphasise the problems of environmental policy decisions and the dilemmas which are a product of a variety of value systems and measurements. The bewildering array of issues demands a framework within which issues may be considered. Pearson's categories of environmental issues goes some way toward providing such a framework. Through the consideration of problems on a spatial and temporal basis and hence through reference to the categories created, at least some of the problems of environmental policy enactment may be better understood. Many of the conflicting evaluations of the environment found in all four chapters are the result of competing interest groups which are approaching the problem from fundamentally different spatial or temporal perspectives. Wind farms in upland areas, while an attractive alternative and an immediate, or short-term, response to energy needs and environmental concerns on a national, regional or global basis, have short- to long-term implications for the local community which cannot be judged alongside the same criteria which make wind farms an attractive alternative at the national or international level. Similarly, the problems of environmental control within the legal framework may appear straightforward at the international level and, through the 'objective' or scientific measurement of water quality, at the local level. With reference to long-held or long-accepted views as enshrined in common law, global long-term and local short-term values will conflict. This theme of spatial and temporally defined issues is again addressed in Chapter 11 with reference to Rome. The local short-term and global long-term evaluation of the environment are encapsulated in the consideration of the local environment in that most international of cities.

The chapters in this section provide an insight into specific policy dilemmas, the complex nature of such dilemmas and the problems which measurement of environmental values raises, rather than solves.

8 Environmental Priorities in Different Development Situations: Electricity, Environment and Development

PETER J. G. PEARSON
Centre for Environmental Technology, Imperial College, London

INTRODUCTION

Discussions about global environmental change raise questions about the value of environmental quality to individual nations as they address domestic issues, and to groups of nations which face transnational environmental problems, such as the enhanced greenhouse effect. The Earth Summit at Rio in 1992, particularly in its discussions on climate change, forestry and biodiversity, illustrated the problems which can arise in reaching workable arrangements for international environmental co-operation and regulation. One cause of the difficulties at Rio was insufficient understanding of the nature and extent of variations in environmental priorities between countries in different development situations.

In the case of the energy sector we can ask what priorities are given to the environmental problems of this sector by countries in different development circumstances, relative to energy policy objectives other than the pursuit of environmental quality, and to environmental problems associated with other sectors. Even within energy, these issues are complex and diverse. Consequently, this chapter focuses mainly on electricity generation. Widely viewed as both a key input to economic growth and a major source of environmental problems, the electricity sector raises important issues about potential conflicts and complementarities between development and environment. Because electricity generation has tended to be the fastest-growing energy sector in developing countries, and also because of its role in global warming, it is necessary to explore the underanalysed differences between developing countries (DCs) and industrialised countries (ICs) in the policy priorities given to environmental quality issues in electricity. It is also important to consider the differences which can arise between the priorities which might be desirable for DCs to pursue and the ways in which actual decisions are influenced or limited by constraints of income and wealth, or by the acceptance of sometimes unnecessarily costly trade-offs between policy objectives, or by inappropriate methods of policy and project evaluation and selection. This chapter briefly attempts these tasks.

Values and the Environment: A Social Science Perspective, Edited by Yvonne Guerrier, Nicholas Alexander, Jonathan Chase and Martin O'Brien. © 1995 John Wiley & Sons Ltd.

ENVIRONMENTAL ISSUES ASSOCIATED WITH ELECTRICITY

There are two main (and related) reasons why environmental issues connected with the electricity sector in DCs have become prominent. First, because of changing views about the urgency and severity of the domestic and international environmental issues associated with electricity. Secondly, as Table 8.1 shows, because not only has DC electricity consumption grown rapidly, with a 1971–89 growth rate of 8.5 per cent per year, it has also grown more than twice as fast as in the ICs (represented in Table 8.1 by the OECD countries, with a growth rate of 3.5 per cent per year). As a consequence, the DC share in world fossil fuel generated consumption doubled from 10 to 20 per cent over this period. Furthermore, Eden's scenarios (also in Table 8.1) imply that the DC share – and also its associated CO_2 emissions – is expected to go on rising well into the next century, as DC growth rates continue to outstrip those of the OECD nations.

Table 8.1   Growth rates of world electricity consumption (per cent per year)

Country groups	Historic growth rates			Eden's Scenarios for 1988–2050	
	1971–80	1980–89	1971–89	'Targeted growth'	'Targeted efficiency'
DCs	9.6	7.3	8.5	4.1	3.5
OECD	4.3	2.7	3.5	1.0	−0.1
World	5.2	3.6	4.4	2.2	1.4

Sources: cols 3–5 calculated from International Energy Agency (1991, pp172–4); cols 5–6 from Eden (1993, Table 8, p235).

For the period of the 1990s, a World Bank study of electricity utility expansion plans in 70 DCs (Moore and Smith 1990), indicated that electricity demand was expected to grow at an average rate of 6.6 per cent per year for 1989–99, raising installed generating capacity by more than 80 per cent, from 471GW to 855GW. In terms of electricity supply, coal is expected to provide almost half, with hydro providing a little less than a third. Coal's contribution to DC electricity supply is expected almost to double in volume, bringing additional problems of local and global pollution. However, projected increases in new generating capacity and supply do need to be treated cautiously because of the severe funding difficulties experienced recently by many poorly performing DC utilities (Barnett 1992; Schramm 1993; World Bank 1993). Nevertheless, even in the face of these well-known difficulties, there seems little doubt that DC electricity supplies – and their associated environmental problems – are likely to continue for some time to come to increase at significant positive annual growth rates well in excess of those of the ICs.

Nine major areas of environmental concern related to electricity generation can be identified (OECD/IEA 1991): major environmental accidents, water pollution, land use and siting impact, radiation and radioactivity, solid waste disposal, hazardous air pollutants, ambient air quality, acid deposition, and global climate change. (See *IEE Proceedings-A*, **140**, 1, January 1993, for a selection of papers which examine various aspects of the environmental impacts of electricity generation.) Given the expected continued growth in electricity generation over the next few decades, it is clear that in the absence of major changes in pricing, management and technology, the environmental impacts associated with electricity are also likely to grow rapidly; a number of electricity supply and air pollution scenarios for the period 1990–2030 were prepared by Anderson and Cavendish (1992) for the *World Development Report 1992* (World Bank 1992). The 'unchanged practices' scenario, in which pollution abatement technologies are not widely used, suggests that the emissions index of three regional and local pollutants rises exponentially at about 6 per cent per year, with the result that DC emissions increase more than fourfold between 1990 and 2010 and tenfold over the 40 years between 1990 and 2030. For global pollution, given the projected role of fossil fuels, especially coal – the most carbon intensive of these fuels – in future DC electricity generation, it is not surprising that studies show DC CO_2 emissions and shares in global emissions rising rapidly well into the next century (Pearson 1992; 1993).

ELECTRICITY, ECONOMIC DEVELOPMENT AND ENVIRONMENT

It is, of course, dangerous to generalise about DC attitudes to policy issues. Nevertheless, Munasinghe (1992) describes a widely held view in DCs that attempts to control environmental impacts should not jeopardise the critical role of electric power in economic development. Adherents to this view are reluctant to allocate funds to environmental programmes since they fear that such expenditures will compete with funding for the expansion of electricity supplies. However, although it is possible to specify conditions in which investments in electricity will form an essential element in a development strategy, and in which expenditures on environmental quality could crowd out such investments, views of the centrality of electricity's role in development and of the competitive nature of the trade-offs between electricity and environmental quality should not be accepted uncritically. (See Lee and Kim (1993, p42) for a discussion of different views about the relationship between energy supplies and economic growth.)

First, the assignment of high priorities to investments in electricity supply may not always represent the best use of scarce development resources. It is not hard to suggest circumstances where the net social benefits of increasing electricity supplies, particularly in rural areas, may be smaller than those obtainable by devoting funds to other non-electricity development projects – for example, on health, water supply, sanitation, education or transport. Secondly, in recent years there has been much advocacy of the possibilities of adopting 'no regrets' (or 'low regrets') measures which simultaneously enhance energy efficiency and also yield environmental benefits (Grubb 1990; 1991). Nevertheless, in suggesting such complementarity between energy efficiency investments and environmental quality, it is important to be aware of both the conceptual and practical limits to these processes, since enthusiasts for 'no regrets' measures sometimes represent them as a panacea. In particular, if energy efficiency improvements effectively deliver energy services at lower costs per unit to the users, it is likely that in many circumstances energy consumption and pollution emissions will rise rather than fall (Anderson 1993).

It is significant, however, that the optimism of a number of commentators about the possibilities of achieving both more electricity supplies and improvements in environmental quality is based principally on their estimate of the benefits from removing existing policy distortions through improvements in *economic efficiency* rather than in narrowly defined technical *energy efficiency*, where energy efficiency is defined as the output of an energy-using activity per unit of energy input, eg the kWh output of a power station per unit of fuel used. For example, Anderson (1993) asserts that raising economic efficiency in energy production and use in DCs would 'liberate' substantial public and private resources which could be allocated to other environmental services such as water supply, sanitation and health. This kind of argument was also pursued in some detail in the *World Development Report 1992* (World Bank 1992). On the other hand, in practice it has proved far from easy in the past to implement economic efficiency measures in the electricity sector (Schramm 1993). Furthermore, while Wilbanks (1991) praises the cost-effectiveness at the margin of investments in improving the efficiency of electricity service provision, he also questions how much efficiency improvement is likely to be achievable in practice in DCs over the next few decades:

> The main reason is that realizing its full potential will call for wrenching redirections by many industrialized countries, developing countries and multilateral institutions – ranging from energy price reform to major changes in financial and technical assistance policies. (p137)

The above arguments suggest, therefore, that we should be cautious before concluding that the pursuit of environmental quality will necessarily compete with the objective of increasing electricity services. Nevertheless, there *are* circumstances in which countries – particularly poor countries – will find it difficult to prioritise environmental quality measures (especially those with a substantial external benefit to other countries) over investments which increase domestic output and incomes. In the case of the threat of global warming, the present costs of restraining anticipated fossil fuel related greenhouse gas emissions and adapting to the potential impacts of climate change imply disturbing intertemporal trade-offs. As Nordhaus (1991) puts it:

> The fundamental policy question involves how much reduction in consumption society should incur today to slow the consumption damages from climate change in the future.

DCs often argue that they cannot afford to ration their use of the atmosphere as a waste sink for CO_2 from fossil fuel combustion, since to do so would compromise their development plans. At the final meeting of Working Group 3 of the Intergovernmental Panel on Climate Change (IPCC) in 1990, the Chinese delegation is reported to have said: 'if we stabilise emissions, that means we cannot develop the Chinese economy' (*The Environment Digest*, 1990, **35/36**, 8–9). Modelling exercises (discussed in Pearson 1992; 1993) tend to confirm the view that countries with significant reserves of carbon-based fossil fuels, such as China and India with their large coal resources, face especially difficult choices. The aspirations of many DCs make them understandably reluctant to do anything which might jeopardise their plans for short- to medium-term economic growth and development. Moreover, given the uncertainties and the long timescale, many DCs will be tempted to take a relatively optimistic view of the net damage which could be associated with global warming in the latter part of the twenty-first century and beyond. As a result (but also for reasons of strategic bargaining), they will tend to set lower targets for CO_2 emissions control than the already developed ICs – and they are likely also to require compensation and technology transfer to persuade them to adopt the targets of ICs (Pearson 1992; Bates 1993).

Of course, one of the reasons why it has proved difficult to reach accommodations between ICs and DCs on greenhouse gas control is that richer ICs want poorer DCs to accept the trade-offs which appeal to richer countries. However, since both the marginal costs and benefits of pollution abatement, and therefore appropriate optimal abatement levels (where marginal costs equal marginal benefits), can differ significantly between countries at different levels of development, the emission abatement levels which make sense to one country may well be inappropriate for another country. Thus Bates (1993) argues that, whereas the ICs have attained most reasonable goals of development and can afford to substitute environmental protection for further increases in material output, the DCs can be expected to take part in the global effort to control CO_2 only where it does not interfere with their immediate development objectives.

ENVIRONMENTAL PRIORITIES AND PRESSURES FOR CHANGE

It has been suggested above that there have been both internal and external reasons why electricity-related environmental issues have become more prominent. Imran and Barnes (1990) suggest that in the eight countries they examined in their study of energy demand (Brazil, China, India, Indonesia, Malaysia, Pakistan, the Philippines and Thailand), the environmental impacts of energy had little impact on policy until very recently. The study suggests that the dominant priority had been the provision of energy to meet the aims of economic development. Furthermore, the scale of the impacts had been viewed as relatively insignificant in relation to the overall scale of the environment; in many nations environmental externalities have been widely accepted as an unfortunate but necessary price to pay for economic progress.

Official attitudes towards such environmental impacts have begun to alter in a number of DCs. This is partly a reflection of growth in domestic and international public pressure. For example, in the 1980s in particular, many domestic and international groups put increasing pressure on governments and international funding agencies, such as the World Bank, to address more seriously the problems associated with the environmental and social impacts of hydroelectric schemes (Goldsmith and Hildyard 1984; 1986; Dixon et al 1989). Issues over resettlement and compensation had proved especially contentious and had been poorly handled (as they often still are). However, DC governments do not always react warmly to such pressures; a number of them discourage or will not tolerate the free expression of environmental concerns. Nevertheless, because of both rising incomes and perceptions of worsening environmental quality in urban areas and international pressures, it seems not unlikely that in the future domestic environmental impacts associated with power projects may claim higher priority in many countries.

The impacts of external *financial* pressures on the environmental impacts of electric power projects are not easily predictable. MacKerron (1991) discusses the potential effects of recent changes in the World Bank's lending policies (and those of other agencies), which have tended to raise the costs of capital. He concludes that the adoption of a higher (implicit or explicit) cost of capital cannot be said *a priori* to have either a net positive or negative environmental impact; total investment volumes are likely to decline in the face of more costly capital, and it is not necessarily the case that because new projects may be relatively short term in conception they will be more environmentally harmful than the investments they displace. For example, combined-cycle gas turbine (CCGT) generation (which is relatively efficient and 'clean' compared with coal) has begun to be popular, especially among independent power producers. However, this is mainly because of financial reasons, including relatively low front-end construction costs and short construction times.

OBJECTIVES, TRADE-OFFS AND PRIORITIES IN THE ELECTRICITY SECTOR

Environmental priorities have to be weighed against a country's other development objectives and these trade-offs will influence decision making.

MacKerron's suggestion (MacKerron 1991) that in the ICs the electric power sector is perceived to be *the* major contributor to environmental problems, is unlikely to hold true for DCs. DCs encounter a variety of other severe environmental problems, some associated with other energy sectors (including the traditional biomass fuel sector) and others with non-energy sectors. These problems range from the prevention of natural resource degradation (including deforestation, soil erosion and their consequences) to the provision of clean water and sanitation and the disposal of waste.

For electricity generation, it is not hard to see why there might exist significant differences in environmental priorities for countries in different development situations – for example between DCs and richer ICs. In Figure 8.1 environmental issues are classified according to four spatial and three temporal categories.

	Local	National	Regional	Global
Short-Term				
Medium-Term				
Long-Term				

Source: based on Pearson (1993, Table 6, p102).

Figure 8.1   Categories of environmental issue

MacKerron (1991) has called attention to two issues in relation to the spatial categories. The first (which applies also to the temporal categories) is about the difficulties in comparing the relative importance of issues across the categories. One method, of course, is to use money as the measuring rod of individuals' preferences, through the efficiency-based economic evaluation techniques of cost–benefit analysis. However, despite recent developments in techniques for the measurement of the demand for environmental quality (Braden and Kolstad 1991), there are still very serious difficulties of data and measurement relating to both the benefits and the costs of policies to control environmental impacts. These difficulties are much more serious in DCs. (For a fairly optimistic discussion of the possibilities of 'transferring' benefit estimates from ICs to DCs, see Markandya 1993.) Moreover, as Pearce (1993) notes, 'In the real world of political decision-making, priorities are rarely set by reference to measures of costs and benefits' (p35).

The second issue relates to the perception of environmental damage. MacKerron (1991) suggests that environmental issues are not yet high on the agenda in many DCs, and that where they are recognised as significant they tend to focus on local, site-related issues, such as the effects of hydro schemes or of urban air pollution. (See also Lee and Jung (1993, p11) for a discussion of the priorities of different income groups of DCs.) There are fewer national and regional issues (eg nuclear safety and national acid deposition problems) and they do not achieve a high political profile, compared with other problems which are perceived as more urgent. Moreover, the global issues, especially the external effects on other nations, attract little serious policy concern as yet in many DCs.

Thus it can be argued that local environmental impacts are likely to receive a higher relative priority in DCs compared with ICs. (This argument is pursued in more detail in the *World Development Report 1992* (World Bank 1992).) Moreover, lower levels of income and wealth tend to constrain DCs to allocating higher priority to the short- and medium-term impacts than to the long-term impacts. This implies that for DCs impacts located near the top left-hand cell in Figure 8.1, such as the short-term local effects of air pollution or the short-term local disruption associated with building hydro systems, may be given a higher ranking than, say, the long-term global impacts of the enhanced greenhouse effect, located at the bottom right-hand cell of Figure 8.1. (Although hydro schemes can also have medium- and long-term environmental and social consequences, some at considerable distances from the principal site.) On the other hand, ICs might be expected to allocate relatively higher priorities to longer-term problems of a less local nature.

Suppose that we observe a tendency for DCs to choose policies or projects which reflect different (often lower) levels of environmental quality in the electricity sector from those which apply in ICs. Some of the following explanations might apply (Pearson 1991):

1. DCs may have different preferences from those of richer ICs and may apply different (often implicit) weights to their policy criteria.
2. *Implicit* discount rates may be unusually high because of poverty and capital constraints, so that projects/policies associated with high levels of environmental quality are not selected even where they show substantial internal rates of return. Here though, it is important to be aware that there is no unique relationship between high discount rates and environmental deterioration (Pearce et al 1990, p26).
3. In any case the projects/policies may exhibit different internal rates of return from those which might obtain in ICs because:
   (i) the costs and/or benefits of the projects/policies, for a given level of abatement, are different from what they would be for an IC;
   (ii) what is taken into account in gauging benefits is not the 'total economic value' (TEV) of social benefits (including use values, option values and existence values – Pearce et al 1991) but something much narrower, so that the stream of benefits is undervalued;
   (iii) TEV *is* accounted for but external beneficiaries (eg foreign governments) are not thought likely to compensate for the component of the benefits which will accrue to them (say from a reduction in CO_2 emissions), so the project/policy becomes unattractive.

4. Some projects/policies with streams of environmental benefits far in the future – but with front-loaded costs – may exhibit a low present discounted value of net benefits compared with other projects with more front-loaded benefits streams. Consequently these types of environmental projects will not be high on the priority list (in terms of a criterion such as the internal rate of return), especially if the undiscounted net benefit streams are low because of points (i), (ii) or (iii) above. (Lee and Kim (1993) discuss the ways in which front-end cost barriers can inhibit the take-up of efficiency-related projects in DCs.)

Clearly, there are important differences between these possible explanations for DCs appearing to choose different targets for environmental quality from those of countries in different development situations. Furthermore, the policy implications range from relaxing income and capital constraints (eg through aid programmes) to the adoption of different practices and criteria in project/policy selection. However, much more work remains to be done in elucidating the relationships between income and preferences for environmental quality in the context of DCs and energy (Pearce et al 1990). As Kopp (1992) notes:

> Although there is probably little doubt that in developing countries a demand exists for the private and quasi-public services of natural assets, it is uncertain whether a demand exists there for the more intangible services of those assets. There is even greater uncertainty regarding the issue of when in the process of economic development that demand would manifest itself.

CONFLICTS AND TRADE-OFFS BETWEEN POLICY OBJECTIVES

Conflicts between policy objectives are an aspect of energy and environmental policy making everywhere. However, it can be argued that the weighting of policy objectives and constraints does, and sometimes should, differ between ICs and DCs. For example, many DCs appear to place relatively more emphasis on objectives relating to the reduction of poverty, the pursuit of equity and the maintenance of macroeconomic stability rather than on economic efficiency. This could be both because of the greater incidence of absolute poverty in poorer countries, and because the often restricted range and power of their economic and social policy instruments, including social security instruments, tends to make it more difficult to resolve conflicts between the pursuit of different objectives. Thus, countries for which poverty objectives are very important may be faced with, and be prepared to accept, significant losses of efficiency, and even equity, as the price of achieving them. As an example, energy subsidies intended to relieve absolute poverty may address this aim at the price not only of lost efficiency but also equity, in that they tend to favour richer groups even more than poorer groups because the richer groups consume more of the subsidised fuel. This, of course, is one of the reasons such subsidies tend to prove politically difficult to reduce.

On the other hand, it has also to be questioned whether the policies adopted by DC governments and utilities are always the result of a rational process of weighing trade-offs between different policy objectives. As Pearce (1993) comments:

> nor, of course, are policies necessarily chosen on a rational basis from the social standpoint: chance, favouritism, patronage, whim and corruption are just as important. (p35)

Moreover, noting that in DCs, on average, electricity prices are barely more than one-third of supply costs and are half those in industrial countries, the World Bank (1992) argues that:

> The reasons for persistent underpricing are largely institutional . . . . Governments frequently intervene in the day-to-day operations of [electricity] utilities and they worry that price increases will exacerbate inflation. Utility managers and their boards may have little say in pricing or investment decisions. Lack of accountability and transparency leads to poor management, either of the utilities themselves or of the state fuel companies that frequently supply them. (p117)

The issue of policy priorities and trade-offs is evidently both complex and significant. Whereas it can be argued that there are circumstances in which it makes sense for DCs to have different priorities from those of countries at other levels of development, it is also clear that many of the choices which authorities actually make imply some very costly trade-offs. Wasteful trade-offs matter especially for DCs because current inefficiencies often feed into future poverty, by damaging the resource base to which later generations will have access. This does not, of course, mean that IC governments do not also make such choices – rather that it is DCs which can least afford to forgo opportunities to raise living standards (Pearce 1993, p34). Thus, the World Bank is highly critical of the widespread use of energy subsidies, not just in DCs, on the grounds that they distort resource allocations, so creating losses of consumer and producer welfare, and stimulate energy use and associated environmental damage to be greater than they would otherwise be. Consistent with this view, Larsen and Shah (1992, p21) claim that, on their estimates, by increasing electricity prices three countries with low existing energy taxes – two DCs and an IC, Indonesia, India and the USA – could achieve substantial *local* environmental gains in the form of health benefits (eg from reductions of nitrous oxides, carbon monoxides, particulates and sulphur dioxides) which would exceed the efficiency costs of carbon taxes. On the other hand, recommendations focused on the narrow pursuit of economic efficiency in the power sector raise some complex issues which are not always easy to address – the resolution of conflicts between policy objectives in DCs is considerably more problematic than some efficiency advocates have been willing to acknowledge (Barnett 1993).

CONCLUSION

In this chapter, I have sought to show that variations in environmental priorities at different levels of development are more than interesting; they reflect differing aspirations and constraints, and they matter both because of their significant influence on domestic policy priorities and choices, and for the realisation of the potential of international environmental co-operation. The values which countries place on environmental assets and quality are influenced by other policy priorities, and it is here that there also exist significant intercountry differences. Important questions concern differences between priorities which might be desirable in pursuit of appropriate development and the ways in which actual decisions are influenced or limited by constraints of income and wealth, or by inappropriate methods of policy and project evaluation and selection, or by the acceptance of sometimes unnecessarily costly trade-offs between policy objectives. Finally, it is clear that as yet we know far too little about the forces which influence energy-related environmental quality in DCs, both in terms of the supply of environmental quality and the changing demands for it.

ACKNOWLEDGEMENTS

This paper arises out of research carried out during tenure of a UK ESRC Global Environmental Change Research Fellowship (award reference Y 320 27 3053). The paper draws on Pearson (1992; 1993; 1994).

REFERENCES

Anderson, D. (1993) 'Energy Efficiency and the Economics of Pollution Abatement', *Annual Review of Energy and Environment*, Annual Reviews Inc, **18**, 291–318.

Anderson, D. and Cavendish, W. (1992) 'Efficiency and Substitution in Pollution Abatement: Three Case Studies', World Bank Discussion Paper No 186, World Bank, Washington, DC.

Barnett, A. (1992) 'The Financing of Electric Power Projects in Developing Countries', *Energy Policy*, April, 326–34.

Barnett, A. (1993) 'Aid Donor Policies and Power Sector Performance in Developing Countries,' *Energy Policy*, February, 100–12.

Bates, R. (1993) 'The Impact of Economic Policy on Energy and the Environment in Developing Countries', *Annual Review of Energy and the Environment*, Annual Reviews Inc, **18**, 479–506.

Braden, J. and Kolstad, C. (1991) *Measuring the Demand for Environmental Quality*, North Holland, Amsterdam.

Dixon, J., Talbot, L. and Le Moigne, G. (1989) 'Dams and the Environment: Considerations in World Bank Projects', World Bank Technical Paper No 110, World Bank, Washington, DC.

Eden, R. (1993) 'World Energy to 2050: Outline Scenarios for Energy and Electricity', *Energy Policy*, **21**, 3, 231–7.

Goldsmith, E. and Hildyard, N. (1984) *The Social and Environmental Effects of Large Dams, Vol 1*, Wadebridge Ecological Centre, Camelford.

Goldsmith, E. and Hildyard, N. (1986) *The Social and Environmental Effects of Large Dams, Vol 2*, Wadebridge Ecological Centre, Camelford.

Grubb, M. (1990) *Energy Policies and the Greenhouse Effect: Vol. 1 – Policy Appraisal*, RIIA/Dartmouth Publishing Company, Aldershot.

Grubb, M. (1991) *Energy Policies and the Greenhouse Effect: Vol. 2 – Country Studies and Technical Options*, RIIA/Dartmouth Publishing Company, Aldershot.

Imran, M. and Barnes, P. (1990) 'Energy Demand in the Developing Countries', World Bank Staff Commodity Working Paper No 23, World Bank, Washington, DC.

International Energy Agency (1991) *Energy Statistics and Balances of Non-OECD Countries 1988–89*, OECD, Paris.

Kopp, R. (1992) 'The Role of Natural Assets in Economic Development', *Resources*, **106**, Winter, 7–10.

Larsen, B. and Shah, A. (1992) 'Combating the Greenhouse Effect', *Finance & Development*, December, 20–23.

Lee, H. and Jung, T. (1993) 'New Direction of Energy Policy in Asia under Environmental Constraints', Working Paper 9306, Korea Energy Economics Institute, Euiwang Si, April.

Lee, H. and Kim, J. (1993) 'The Role of Technology Transfer in Coping with Global Warming: the Case of Korea', Korea Energy Economics Institute, Euiwang Si, June.

MacKerron, G. (1991) 'Environmental Impacts of Investment Policies in Developing Country Power Systems', in P. Pearson (ed) 'Environment, Energy Efficiency and the Third World', Surrey University Energy Economics Centre Discussion Paper (SEEDS 60), Guildford.

Markandya, A. (1993) 'Air Pollution and Energy Policies: the Role of Environmental Damage Estimation', Working Paper 41.93, Fondazione ENI Enrico Mattei (FEEM), Milan.

Moore, E. and Smith, G. (1990) 'Capital Expenditures for Electric Power in the Developing Countries in the 1990s', Industry and Energy Department Working Paper, Energy Series Paper No 21, World Bank, Washington, DC.

Munasinghe, M. (1992) 'Efficient Management of the Power Sector in Developing Countries', *Energy Policy*, **20**, 2, 94–103.

Nordhaus, W. (1991) 'To Slow or Not to Slow: the Economics of the Greenhouse Effect', *Economic Journal*, **101**, 407, 920–37.

OECD/IEA (1991) *Greenhouse Gas Emissions; the Energy Dimension*, Organisation for Economic Co-operation and Development, Paris.

Pearce, D. (1993) *Economic Values and the Natural World*, Earthscan, London.

Pearce, D., Barbier, E. and Markandya, A. (1990) *Sustainable Development*, Earthscan, London.

Pearce, D., Barbier, E. and Markandya, A. (1991) *Blueprint 2: Greening the World Economy*, Earthscan, London.

Pearson, P. (1991) 'Externalities, Energy and Environmental Control in the Third World', in 'Governance by Legal and Economic Measures', 8th Annual Conference of the European Association for Law and Economics, Copenhagen, August (mimeo).

Pearson, P. (1992) 'Greenhouse Gas Scenarios and Global Warming: the Role of Third World Countries', in G. Bird (ed) *International Aspects of Economic Development*, Academic Press, London, 125–67.

Pearson, P. (1993) 'The Environmental Impacts of Electricity Generation in the Third World', *IEE Proceedings-A, Science, Measurement & Technology*, **140**, 1, 100–108.

Pearson, P. (1994) 'Environmental Quality, Electricity and Development: A Review of the Issues', Department of Economics Discussion Paper No 94–09, University of Birmingham, Birmingham, May.

Schramm, G. (1993) 'Issues and Problems in the Power Sectors of Developing Countries', *Energy Policy*, July, 735–47.

Wilbanks, T. (1991) 'The Outlook for Electricity Efficiency Improvements in Developing Countries', in 'Electricity and the Environment', Background Papers for a Senior Expert Symposium in Helsinki, May, IAEA-TECDOC-624, Vienna, 121–45.

World Bank (1992) *World Development Report 1992*, Oxford University Press, Oxford.

World Bank (1993) 'The World Bank's Role in the Electric Power Sector', *World Bank Policy Paper*, World Bank, Washington, DC.

9 Wind Farms: A Case of Conflicting Values

MERYLYN McKENZIE HEDGER
Institute of Earth Studies, University of Wales, Aberystwyth

INTRODUCTION

The development of renewable energy sources was given an impetus with the oil price hikes of the 1970s; indeed, until the early 1980s their role was principally seen as substituting for fossil fuels as these became scarcer. Currently, they have increased prominence because of the perceived need to control atmospheric emissions of greenhouse and other gases and substances (UNCED 1992). It is recognised that strategies must be deployed which will be based on efficiency in energy production, transmission and consumption, and that there will be a growing reliance on environmentally sound energy systems, particularly new and renewable sources of energy (UNCED 1992). Furthermore, their proponents argue that if the world economy is to expand to meet the aspirations of countries around the globe, energy demand is likely to increase and, given adequate support, it is the renewable energy technologies which should be used as they can meet much of the growing demand at prices lower than those usually forecast for conventional energy (Johansson et al 1993). Apart from their advantages for abatement of global warming and air pollution, their supporters argue that they have other benefits which are not usually captured in standard economic accounts, such as social and economic development in rural areas at local level, fuel supply diversity at international level and the reduction of the risks of nuclear weapons proliferation (Johansson et al 1993).

However, there are problems in capturing natural ambient energy. Most of the time, natural energy flows are not sufficiently concentrated to be converted rapidly into useful quantities of fuel or energy (Patterson 1990). Natural energy flows vary widely and are unpredictable, and to convert them reliably and efficiently into fuel or electricity is a major challenge. Even when technological problems appear to have been resolved, major problems can arise because to capture ambient energy on a sufficient scale to replace significant amounts of fossil fuel or nuclear capacity renewables may require larger land takes, land which is already in other uses (OECD 1988).

Values and the Environment: A Social Science Perspective, Edited by Yvonne Guerrier, Nicholas Alexander, Jonathan Chase and Martin O'Brien. © 1995 John Wiley & Sons Ltd.

The development of wind farms in the uplands of England and Wales provides an interesting case of a conflict over land use, the more so because it is being conducted not only between competing demands for land use for economic reasons but also between environmentalists with different visions. This chapter explores the conflict which has been provoked, interprets the viewpoints of the opposing groups, and considers the context in which the confrontation takes place.

DEVELOPMENT OF RENEWABLE ENERGY IN THE UK

Experience in the UK has confirmed that no transition to renewables readily occurs if market conditions remain unchanged and there is no public policy framework which is conducive to the development of renewable energy. Research programmes were supported from the 1970s but only at modest levels, as their large-scale exploitation was thought to lie 'well into the future' and their contribution would only begin to build up usefully in the first quarter of the next century (Cmnd 7101 1978). It was the government's White Paper on the environment in 1990 which set the first target of a new renewable energy electricity generating capacity of 1000MW in 1990 (Cm 1200 1990). The Coal Review accepted the recommendation of the Renewable Energy Advisory Group and increased the 1000MW target of installed renewable capacity by the year 2000 to 1500MW (DTI 1993a; 1993b). The latest official statement confirms this target and indicates that it is government policy to stimulate the development of renewable energy wherever it has the prospect of being economically attractive and environmentally acceptable in order to achieve supply diversity, emission reduction and the encouragement of internationally competitive industries (DTI 1994a, b).

The move from limited research projects to commercial reality has come with the introduction of a supportive subsidy, and a new legislative and institutional framework. This new system arose, not principally because of a desire to further the development of renewable energy technologies, but because of a need to address the problem of nuclear power during the privatisation of the electricity industry (Ilex 1991). A new statutory obligation, the Non-Fossil Fuel Obligation (NFFO), was introduced on the Regional Electricity Companies to contract for a specified minimum of non-fossil fuelled generating capacity (DTI 1993b). To provide a 'green sweetener', this also included renewable energy sources. The cost of subsidies is financed by a levy of approximately 10 per cent on consumers' bills. In 1992/3 this raised £1348 million, but the sums diverted to renewables are small, 3 per cent initially then increasing to 6 per cent in 1993/4. Owing to the linkage with nuclear power, the European Commission initially insisted that the subsidy should end after eight years (in 1998) – but the end limit was lifted in 1993 for the next orders for renewable energy. Two orders were announced in 1991, 1992, an announcement on bids for the 1995 order will be made in November, and there will be other orders in 1995 and 1997 (DTI

1993c). The 1991 and 1992 orders did not cover Scotland or Northern Ireland; separate orders for them have been announced.

The development of renewable energy in the UK has, however, been constrained in a number of ways by the operation of the NFFO mechanism. It has had the effect of supporting those technologies which are more technically mature. Wind energy falls into this category, largely owing to support by other governments, particularly the Carter administration in the USA following the energy crises of the 1970s. The initial 1998 cut-off mechanism could have worked against wind energy because of its capital intensiveness. However, government has clearly been determined to support the emerging industry because wind-generated electricity has been commissioned, despite being the most expensive: 11p/kWh compared to 5.7p/kWh for landfill gas and 6.0p/kWh for hydroelectricity in the 1992 order. Nevertheless, there have been commercial imperatives operating and the higher wind speed sites (over 7.5m/sec) are the most attractive. At the public inquiry on the wind farm at Mynydd y Cemmaes, Powys, the developers stated that at sites of 7.5m/sec wind speed costs were 10p/kWh, but at sites of 5.5m/sec costs were 26p/kWh. It would appear from the statement by the Minister for Energy on the Third Renewables Order on 21 July 1993 that a 15–20 year limit may operate in the next tranches. The Minister for Energy has made it clear that in the current order there should be convergence with commercial prices (DTI 1993c).

So far, only a small number of wind farms are working in England and Wales. In June 1994 23 had been constructed or were being completed. This in itself has been hailed as an impressive achievement in three years (Hodgson 1993). There is evidence, however, of an explosion of developer interest. In March 1994, the Minister for Energy reported that as many as 230 applications had been received for the 1994 NFFO round (DTI 1994a). This level of interest reflects the immensity of the resource, identified as being the best in Europe (DTI 1994b). Furthermore, it reflects the growing perception that 'wind energy is on the verge of commercial exploitation' (DOE and Welsh Office 1993), so it is the high wind speed areas which will continue to be the most attractive to develop.

The problem arises because of the siting requirements for wind farms for open exposed areas. They involve turbines which are tall, man-made structures with noise-emitting moving components, requiring deep foundations and to which permanent, wide access roads are required. The potential impacts of a large-scale programme of wind energy have been identified for many years. In the Consultative Document on Energy Policy issued in 1978, for example, it was stated that:

> A large number of wind generators, sited in prominent positions, would be required to yield significant quantities of energy and they would have a substantial visual impact. They are likely to be fairly noisy. (Cmnd 7101 1978)

The areas of high wind speed, the upland exposed areas, are generally those which were not favoured for permanent settlement. The consequent lack of development pressures in these areas meant that their remoteness, wildness and natural beauty have been valued, as the UK has a long history of concern for landscape protection and recognition of the need to protect remote and tranquil areas. The value of 'natural beauty' and 'unspoilt countryside' was perceived from the early nineteenth century with rapid change in towns and cities. Wordsworth's elitist approach was replaced during the century by a broad movement and several organisations were formed before 1914, such as RSPB (Royal Society for the Protection of Birds), National Trust (for places of historic interest and natural beauty) and Open Spaces Society (1865). With new threats to the countryside after 1918 from afforestation and suburbanisation, other organisations were formed: Council for the Protection for Rural England (CPRE), Campaign for the Protection for Rural Wales (CPRW), Ramblers' Association, and Youth Hostels Association (Blunden and Curry 1989). These concerns led eventually, after the Second World War, to the institution of designations for landscape and nature conservation protection. Through the planning system, these and other upland areas have received protection from developments apart from agriculture and forestry.

More recent environmental groups such as Friends of the Earth and Greenpeace, both founded in 1971, have worked to entirely different agendas, focusing on systemic pollution of environmental media and biological and chemical damage to ecosystems. For these groups, wind power is viewed as having environmental benefits as a source of clean energy. Many in this grouping see turbines as beautiful and can readily accept the visual impacts as being positive symbols of a new environmentally aware age. They claim that 'the true guardians of our countryside' will give support to appropriate wind power projects (*Planning*, 10 June 1994).

With such basic differences in approach, it is therefore not surprising that the rapid implementation of a programme of a large number of wind generators with substantial visual and noise impacts is creating controversy. Conflicts have arisen through proposals to site individual wind farms at local level when planning applications have been made, and in some cases during and after construction. These disputes have been picked up by regional news media and thence the national press, particularly when well-known names have become involved. Opposition by Sir Bernard Ashley (co-founder of the Laura Ashley shops) to a wind farm in Mid Wales funded by the Body Shop led to headlines like 'Battle of the Green Giants' (*Western Mail,* 13 September 1993). Sixty of the literary great and good, led by poet Ted Hughes, a local resident, signed a letter to the *Times Literary Supplement* (19 February 1994) about the proposed wind farm at Flaight Hill and its 'despoliation' of Brontë Country, which led to a *Times* editorial in support (19 February 1994). *The Economist* has written about 'windfarms – a new way to rape the countryside' (22 January 1994). An article by Matthew Engel in *The Guardian* (26 March 1994) made an analogy between the development of wind power and the Vietnam War (ie wind power was destroying what it was purporting to save). Needless to say the pro lobby has been fighting back, starting to blame the media for distorting the situation and using bogus arguments (Harper 1994), asserting that proponents are pro-nuclear (*Cambrian News*, 10 December 1994) and misinformation by a handful of activists (*Planning*, 10 June 1994).

The conflicts are not just a result of NIMBY attitudes, nor just about competing rationales about priorities for environmental protection, but are also motivated by the protection of vested, economic interests. There are two broad groupings of opinion and the rationales and composition of the two opposing viewpoints will now be considered.

SUPPORTERS

A new alliance is developing in support of wind farms, comprising environmental groups, local authorities keen to find any economic opportunity for remote, rural areas, and economically depressed farmers. The supporters will be considered in five broad groups: developers; manufacturers; landowners; the local community; and the 'new' environmental movement.

Developers and Manufacturers

As indicated, the national economic interest has recently emerged as an issue and the government sees encouragement of the national industry as a main objective of its policy (DTI 1993b), no doubt reflecting concern that so many of the developments have been using technology from overseas countries such as Denmark, whose governments have been actively supporting wind energy development for some time. It has been the manufacturers which have fought on this issue, both individually and through their professional organisation the British Wind Energy Association (BWEA). This lobby, however, is not cohesive 'because the needs of big project developers are very different from those of private individuals' (*Windpower Monthly*, May 1993) and apparently a 'subgroup has emerged within the BWEA as a trade lobby for the larger suppliers' (NATTA 1993). The British industry includes large-scale, long-established firms which are diversifying, notably British Aerospace, Taylor Woodrow and McAlpine. The new industry has also provided opportunities for new entrants, including new ecobusinesses. Among the developer companies in the UK at present, there are several which involve technical specialists who would regard themselves as committed environmentalists and who have been working to promote the cause of renewable energy for some time. Thus 'alternative' enterprise is now becoming mainstream. For example, the developing company which achieved most success during the second NFFO round with 13 approvals from 13 applications, Ecogen, has roots in the National Centre for Alternative Technology, Machynlleth, Powys and the Cornwall Wind Energy Project.

Landowners

With overcapacity in food production in the agricultural industry, and the prospect that agricultural support will decline in the future, big surpluses of land have been identified (Green 1993). Farmers are looking for possibilities for diversification. They now have a strong economic motivation for seeking wind turbines on their land as they will gain between £1000 and £2000 a year for each turbine (Ellis 1992), which is comparatively a very profitable crop, particularly as farm incomes have been declining significantly until recently. Farm unions in Wales have been active in support and have organised conferences covering wind energy. Apart from the direct economic motivation, in both Wales and Cornwall early wind farms can be linked to landowners who had longstanding interests in renewable energy – they are now acting as consultants to other farmers.

The Local Community

It is difficult to generalise about attitudes to wind farms at local community level on the basis of responses to date and knowledge from special surveys. Communities are not cohesive units and opinion within them invariably falls into two camps. Wind farms have stimulated the usual anti-development responses, but a striking feature of their progress so far is that they have also managed to inspire local support in some rural areas. Surveys of local attitudes towards wind farms in south west Cornwall and Wales have revealed generally positive attitudes towards them. It would seem that many people in some areas do not regard them as visual eyesores and their development, once completed, is perceived as causing little disruption to living patterns (Blandford Associates 1993; ETSU 1993).

The origins of these favourable attitudes probably lie in economic conditions and environmental perceptions. The factors which have led sites to be desirable locations for wind farms also tend to mean that they are economically depressed areas with few other locational advantages for economic development. The local community often perceives direct and indirect economic benefit from jobs during construction and after with ecotourism. Concern to safeguard jobs for a local engineering firm materially affected the consideration of a planning application for a wind farm in Mid Wales recently (*Western Mail*, 11 June 1993). Developers see the provision of funds to the local community as a leverage point to improve acceptability (*Windpower Monthly*, November 1993). As the development is a novelty, the area also can be expected to gain significance and this influences the attitudes of those who could not expect to gain economically. Local attitudes would also appear to reflect an increased environmental consciousness of environmental issues such as atmospheric pollution and problems with nuclear waste. In Wales, wind farms initially benefited from concerns over safety at Trawsfynydd nuclear power station and political fallout from the Chernobyl effect. (Sales of sheep are still restricted in some parts of North Wales, and the closure of Trawsfynydd nuclear power station was announced in July 1993 because of the costs of safety measures.) The Chernobyl effect is widely known and understood, particularly by leaders in the agricultural community who see wind energy as being clean and non-polluting.

The New Environmental Movement

Organisations such as Friends of the Earth and Greenpeace see wind power as an integral part of a strategy in a new environmentally attuned energy policy to reduce both acid rain and global warming This lobby talks about already degraded upland grassland (FOE Cymru 1994), blanket conifer and wet desert, and 'featureless and visually boring upland landscapes' (Backhouse 1994). Calculations are made as to how many tonnes of carbon dioxide emissions each wind farm can save (BWEA 1992). Advocates of wind power also point to its flexibility: in four years in the USA 1000MW of capacity was installed, equivalent to a nuclear power station which would have taken eight or nine years to build, after planning and approval stages (World Bank 1986). It is also argued (controversially) that wind farms have a small land take, as little as 1 to 1.5 per cent of the total involved in a site. This lobby does admit that there will be landscape impacts, but it is considered that these are not sufficient to offset the advantages; it is argued that they are reversible as the turbines can be removed at the end of their useful life.

THE OPPOSITION

Opposition to wind farms extends even to their name in that 'it implies an inherent compatibility with the rural environment', and this lobby prefers names such as 'windpower stations' (CPRE and CPRW 1991) or 'wind turbine power stations' (CCW 1993). This lobby has a greater coherence and revolves around the landscape protection rationale. As more wind farm planning applications emerge, there is evidence of increased polarisation. There are three main groupings in this camp: local residents; statutory agencies responsible for landscape protection – Countryside Commission (England) and Countryside Council for Wales; the 'old' environmental groups – Council for the Protection of Rural England (CPRE) and Campaign for the Protection of Rural Wales (CPRW), Open Spaces Society (wind energy on common land), Council for National Parks (wind energy in National Parks) and Ramblers' Association.

Local Residents

As indicated, there seems to be a wide variation in reactions to proposals for wind farms. Opposition seems more likely in England than in Wales, as a far higher proportion of sites included in the first and second NFFO rounds have been refused, possibly because of higher population densities as opposition tends to be greater where turbines are close to houses. The visual impacts are a major source of concern and noise from the turbines has created anxiety, particularly as those residential locations had been sought for their peace and quiet. Construction periods can be a source of major aggravation as massive engineering works are required and large equipment needs to be transported on narrow roads. Much bitterness has been generated particularly where local residents have experienced greater impacts, such as noise, than they had been led to believe would occur. One battle-worn family is apparently moving house to escape a nearby development (*Daily Telegraph,* 31 May 1994). There has also been concern about noise impacts in urban areas such as Shoreham Harbour (*Surveyor,* 26 May 1994). A gradual awakening of opposition throughout England and Wales has led to the foundation of a new organisation, Country Guardian, which seeks to act as an information network and advice centre to groups and individuals opposing wind farms. One of its main aims is 'to oppose wind farms as being unacceptably damaging to landscapes, economically unsound and irrelevant to energy issues'. It also has aims relating to agriculture and forestry.

Statutory Agencies and 'Old' Environmental Groups

All these groups are concerned with landscape conservation. The statutory agencies, Countryside Commission and Countryside Council for Wales, have as one of their principal functions the responsibility to promote and keep under review the conservation and enhancement of natural beauty in the countryside, designated areas and elsewhere. These groups do not dispute the benefits of renewable energy and the need for the development of its technologies, nor indeed do they oppose all wind farm developments. Where they differ from the wind farm supporters is in arguing that the negative effects, principally the landscape impacts, are too great to offset the benefits of wind farm development on the scale which would be required if it was to make a significant contribution to UK energy supply or the reduction of carbon dioxide emissions. They perceive that:

> There is a danger that the development of renewable energy sources will be used as a 'technical fix' for what is essentially a social, political and economic problem, namely the profligate and unsustainable use of energy. (CPRE and CPRW 1991)

This lobby has therefore been particularly concerned from an early stage about the absence of a clear national energy policy and a strategic approach on wind power (Countryside Commission 1992; Countryside Council for Wales 1993). It argues that capital would be better invested in securing energy efficiency improvements. It asserts that far from being benign, 'farm' activities, wind power stations are industrial and commercial development (CPRE 1991). Calculations are made as to the overall number of turbines required to replace significant quantities of fossil fuels and land areas likely to be affected (drawing boundaries around the whole site). This lobby expects irreversible effects as a consequence of the provision of service roads and consequent changes to land use patterns (CCW 1993). It argues that assessment for green development would be less stringent than for other power stations or developments (CPRE and CPRW 1991).

Battles have been fought on the siting of telecommunications equipment such as single radio masts in prominent locations (Countryside Commission 1985, 1988). Careful attention has been given to the siting of pylons – often deliberately in valleys along route corridors after a lengthy selection and consultation process. The statutory agencies have been lobbying for the past 20 years for changes to agricultural and forestry policy and have met with some success, culminating in the adoption of the agri-environmental zonal programmes by the government in 1993, so they perceive the government's approach to wind power as going into reverse on landscape policy.

THE CONTEXT OF DISPUTE

There has been increased polarisation in the debate. Two environmental groups, Ramblers' Association and CPRW, have hardened their position. In November 1993, CPRW called for a moratorium on payments to wind farms through the NFFO until substantial results had been achieved by reducing energy demand through conservation (CPRW 1993). The Ramblers' Association has now also called for a moratorium on subsidies pending a re-evaluation of the benefits of wind power and the development of a coherent national energy policy (Ramblers' Association 1994). According to its national director, its members reversed their policy of support for wind farms because experience had shown them to be noisy and visually intrusive (*The Guardian*, 26 March 1994). The Wales Tourist Board has identified the spread of wind farms as a potential threat to the well-being of the tourist industry, for which the environment is one of the industry's greatest assets creating 95 000 jobs (Wales Tourist Board 1994). There have been increasing calls for a strategic approach to policy (Association of County Councils 1993; Assembly of Welsh Counties 1993; Welsh Wind Energy Forum 1993). In view of the high profile of the debate in Wales, the Welsh Affairs Committee undertook a very thorough inquiry into wind energy during 1994 (Welsh Affairs Committee 1994).

There has been a growing fear in the industry that vehement opposition to developments, particularly on visual grounds, represents a major threat to future prospects (*Planning*, 10 June 1994). To counter this Friends of the Earth, which has consistently tried to overcome obstacles to wind energy developments (FOE 1990; 1991), has produced new guidelines intended to provide developers and planners with outline standards and practices for project siting, environmental assessment and noise targets and a checklist for minimum development practice (FOE 1994). The British Wind Energy Association is also producing a good practice blueprint designed to respond to widespread concern (*Western Mail*, 15 June 1994). Both these responses are predicated on the belief that with better information from developers and better assessments of public opinion as a whole the 'vociferous anti-wind power minority' can be silenced (*Planning*, 10 June 1994).

This approach seems to under-represent the rationale and attitudes behind the opposition. Landscape protection and countryside preservation have probably always been a minority interest, but nevertheless have been pursued in the face of powerful opposing interests at many times. Wind power is not being developed on a community basis in the UK at present, and until other alternatives such as energy efficiency have been exhausted, it is unlikely to stimulate a large groundswell of support as a main greenhouse strategy.

The major problem is that there is no institutional framework or political will to create a mechanism in which the merits or otherwise of a large-scale wind energy programme can be fully assessed, using, for example, strategic environmental appraisal. At present, decisions are being made on a site-by-site basis through the 'will secure' test by the Office of Electricity Regulation (OFFER) and determination of the planning application by the local planning authorities. (Privatisation gave new power to LPAs to determine electricity-generating plant. Plant over 50MW goes to the Minister and the proposal for the 267 turbine wind farm at Kielder Forest was handled by the DTI.) These two assessments are undertaken entirely independently of each other, principally because the first is conducted in secret for reasons of commercial confidentiality. Some schemes have received approval under one assessment but not the other. There has been concern that planning authorities are under some pressure to approve schemes which have passed the 'will secure' test (McKenzie Hedger 1995). Not only is there a lack of synchronisation between the two procedures at present, but neither allows for systematic measurement of the macroeconomic and environmental economic impacts of wind power technology and thus provides a hard, quantifiable dimension to the debate. Methodologies for the economic valuation of landscape are far from satisfactory and costing the environmental impacts presents conceptual problems of definition:

1. It is necessary to decide at what level the visual impacts are to be evaluated. Should the perception of visual pollution/satisfaction be on site, at local level, the sub-regional or national level and for whom? How are accumulative effects to be measured?
2. Although noise is the easiest impact to measure, there are new challenges because wind farms produce quite distinct types of noise, which are regarded as intrusive not by normal noise standards but because they occur in locations sought for their peace.

CONCLUSION

Therefore, does the development of wind farms represent:

1. At last, the opportunity that renewables require to achieve technological development and credibility?
2. A useful contribution to CO_2 reductions?
3. A development with limited landscape effects which are reversible, and some say even landscape-enhancing effects?
4. A means of providing economic benefit in rural communities?

or:

5. A development which is occurring as a result of a distorted market mechanism which has been little thought out and which will have irreversible effects, on the scarce, finite resource of upland and wild, unspoiled areas?
6. A development whose impact on CO_2 reduction is negligible unless constructed on a massive scale, and the resources for which would be better invested in energy efficiency technologies?

No one is considering which argument has more merit or how they can be reconciled. At present, conflict is intensifying, while the government adheres to its approach that a bottom-up, site-by-site approach within the planning system will prevail, despite the variations in decision making within local authorities and requests from all quarters for strategy guidelines.

REFERENCES

Association of County Councils (ACC) (1993) *Energy Policy: Delivering a Sustainable Energy Policy*, Association of County Councils, CC Publications, London.

Assembly of Welsh Counties (AWC) (1993) *Strategic Planning Guidance in Wales, Overview Report*, Submission to Secretary of State for Wales.

Backhouse, M. (1994) 'Windfarms in Montgomeryshire', *Planning in Wales*, Royal Town Planning Institute, South Wales Branch, Cardiff, May, p7.

Blandford Associates (1993) *Public Attitudes to Wind Power Generating Stations in Rural Wales*, Report to Countryside Council for Wales, Chris Blandford Associates, Cardiff with School of Sociology and Social Policy, University College of North Wales, Bangor.

Blunden, J. and Curry, N. (1989) *A People's Charter?*, Countryside Commission, HMSO, London.

British Wind Energy Association (BWEA) (1992) *Wind Energy and the Environment*, Factsheet 14, BWEA, London.

Campaign for the Protection of Rural Wales (CPRW) (1993) *Wind Power Policy*, November, Campaign for the Protection of Rural Wales, Welshpool, Powys.

Cm 1200 (1990) *This Common Inheritance*, HMSO, London.

Cmnd 7101 Department of Energy (1978) *Energy Policy: a Consultative Document*, HMSO, London.

Council for the Protection of Rural England (CPRE) (1991) *Comments on Draft Planning Policy Guidance Note on Renewable Energy*, CRPE, London.

Council for the Protection of Rural England (CPRE) and Campaign for the Protection of Rural Wales (CPRW) (1991) Comments by CPRE and CPRW to the Department of Energy's Renewable Energy Advisory Group, October.

Countryside Commission (CC) (1985) *Annual Report of the Countryside Commission 1984–1985*, Countryside Commission, Cheltenham, Glos.

Countryside Commission (CC) (1988) *Annual Report of the Countryside Commission 1987–1988*, Countryside Commission, Cheltenham, Glos.

Countryside Commission (CC) (1992) 'Evidence to House of Commons Energy Committee', *Renewable Energy*, **11**, 4th Report, HMSO, London.

Countryside Council for Wales (CCW) (1993) *Wind Turbine Power Stations*, Policy Document, Countryside Council for Wales, Bangor, Gwynedd.

Department of the Environment (DOE) and Welsh Office (1993) *Renewable Energy*, Planning Policy Guidance Note 22, February.

Department of Trade and Industry (DTI) (1993a) *The Prospects for Coal, Conclusions of the Government's Coal Review*, Cm 2235, HMSO, London.

Department of Trade and Industry (DTI) (1993b) *Renewable Energy, Planning for the Future: A guide for local authorities*, DTI, London.

Department of Trade and Industry (DTI) (1993c) *Renewable Energy Bulletin No 5: Information on the Non-Fossil Fuel Obligation for Generators of Electricity from Renewable Energy Sources*, REB5 DTI, October.

Department of Trade and Industry (DTI) (1994a) 'Tim Eggar Spells out Challenge for Wind Energy', *News Notice P/94/143*, Department of Trade and Industry, June 11.

Department of Trade and Industry (DTI) (1994b) *Energy Paper 62*, March, HMSO, London.

Ellis, G. (1992) 'Wind Energy in Farming – A Balanced View', Notes from Presentations, Wind Farming in Wales, NFU Conference, Welsh Agricultural College, Aberystwyth, April 7.

Energy Technology Support Unit (ETSU) (1993) *Attitudes Towards Wind Power: A Survey of Opinion in Cornwall and Devon*, W/13/00354/038/REP, ETSU, London.

Friends of the Earth (FOE) (1990) *Developing Wind Energy for the UK*, Friends of the Earth, London.

Friends of the Earth (FOE) (1991) *Removing the Windbrakes: Wind Energy, the Landscape and the Government*, Friends of the Earth, London.

Friends of the Earth (FOE) (1994) *Planning for Wind Power*, Friends of the Earth, London.

Friends of the Earth Cymru (1994) 'Letter', *Cambrian News*, Aberystwyth, Dyfed, December 24.

Green, B. (1993) 'Agricultural Over Capacity: Strategies for More Sustainable Land Uses', Report of Seminar, The Non-Food Uses of Crops and Land, The Royal Society, London, December 20.

Harper, M. (1994) 'No Wind Situation', *New Statesman and Society*, March 11, London, pp30–31.

Hodgson, S. (1993) 'Wind Passes the 100 MW Milestone', *Review: The Magazine of Renewable Energy*, **21**, December, DTI, pp5–7.

Ilex (1991) *The Non-Fossil Fuel Obligation*, A Report to the Countryside Council for Wales, Ilex Associates, Oxford.

Johansson, T., Kelly, H., Reddy, A. and Williams, R. (1993) *Renewable Energy, Sources for Fuels and Electricity*, Earthscan, London and Island Press, Washington, DC.

McKenzie Hedger, M. (1995) 'Windpower: Challenges for Planning Policy', *Land Use Policy*, January.

Network for Alternative Technology and Technology Assessment (NATTA) (1993) *Renew: Natta Newsletter,* **85** Sep/Oct, EERU, The Open University, Milton Keynes, p4.

OECD (1988) *Environmental Impacts of Renewable Energy: the OECD Compass Project*, OECD, Paris.

Patterson, W. (1990) *The Energy Alternative*, Macdonald, London.

Ramblers' Association (RA) (1994) Motion Approved by National Conference, March 26, Ramblers' Association, London.

United Nations Conference on Environment and Development (UNCED) (1992) *Agenda 21*, Chapter 9, Protection of the Atmosphere, Section B. Promoting Sustainable Development, Part A, Energy Development, Efficiency and Consumption, UNCED, New York.

Wales Tourist Board (1994) *Tourism 2000*, Wales Tourist Board, Cardiff.

Welsh Affairs Committee (1994) *Wind Energy*, Report & Minutes of Evidence, Vols I and II, House of Commons Session 1993–4, July.

Welsh Wind Energy Forum (WWEF) (1993) Letter to Secretary of State for Wales, Secretary of State for the Environment and President of the Board of Trade, November 1993, from Mr M. Backhouse, Chairman, WWEF, Montgomery District Council.

World Bank (1986) 'Guidelines for Assessing Wind Energy Potential', World Bank Energy Department, Energy Paper No 34, World Bank, Washington, DC.

10 Orthodoxy and the Judiciary's Approach to Environmental Impairment: Legal Foresight and Environmental Myopia

MICHAEL JOHN DOUPÉ
Department of Law, University of Lancaster

INTRODUCTION

The observations in this chapter find their origins in the title of the United Kingdom Environmental Law Association's fourth annual Garner Lecture in 1991 (*Journal of Environmental Law*, 1992). The lecture was disingenuously entitled 'Are the Judiciary Environmentally Myopic?' but in the writer's view the lecture signally failed to address this title and could have been more accurately entitled 'The faculties of the Judiciary are fine – it's the legal system and the laws which are failing!' The lecture was delivered by Lord Woolf, one of the newest 'Law Lords' and widely regarded as a liberal judge. This chapter attempts to address the former title by reference to an environmentally significant case, Cambridge Water Authority v Eastern Counties Leather plc, which has wound its way through the gamut of the English civil courts system. This has entailed a first hearing in the High Court, a subsequent appeal to the Court of Appeal and a final appeal to the Appellate Committee of the House of Lords.

THE LEGAL ARGUMENTS

The case concerns the 300-year-old tanning industry situated in the Cambridgeshire village of Sawston, and in particular the environmental consequences of the careless use of organochlorines within that industry. The case was brought by Cambridge Water Company against Eastern Counties Leather plc as a result of the latter's pollution of the former's borehole from which water was abstracted for public supply.

In 1976 when Cambridge Water Company started to abstract, the water was 'wholesome' in accordance with current water quality standards set out in, or under the authority of, the Water Acts 1945 and 1973. However, a European

Values and the Environment: A Social Science Perspective, Edited by Yvonne Guerrier, Nicholas Alexander, Jonathan Chase and Martin O'Brien. © 1995 John Wiley & Sons Ltd.

Communities Directive in 1980 (80/778/EEC2) introduced new and higher standards for drinking water quality which, although aimed at protection of human health, had the additional effect of protecting the environment in that drinking water sources must be sufficiently free from contamination to allow inexpensive water treatment. The new standards were to be met by 17 July 1985 and the technology to carry out the testing which was required by the Directive was available by 1983, at which time it was discovered that the Water Company's borehole was contaminated with organochlorine solvents. Cambridge Water Company was forced to close down the borehole and, at a cost of £1 million, to relocate its abstraction operation. Cambridge Water Company identified the likely sources of the contamination which had been caused by spillages of the solvent, and in 1985 issued a writ claiming damages and an injunction against Eastern Counties Leather and another tanning company (this other company was subsequently shown not to be legally liable).

It may be considered surprising that the claim against Eastern Counties Leather is not based on any legislative provisions of either the UK Parliament or the European Community. This is particularly so in the light of the major reorganisation of the water industry in the late 1980s and the enactment of substantial amounts of legislation to achieve and regulate this basic human need, and also the substantial amounts of European Community legislation which are aimed at environmental protection and which were the cause of Cambridge Water Company's problems. The simple reason is that there is no such basis for legal action. Therefore Cambridge Water Company only had recourse to remedies under what is often described as the greatest creation of the English legal system, namely the common law. The common law of England has its origins in the decisions made by the King's judges in the twelfth century and continues in an unbroken line of continuing development to the present day. In essence this is judge-made law, in line with the precedent of previous decisions, but capable of further development as the needs of 'justice' may dictate. Clearly this body of law is capable of considerable manipulation in the hands of the judiciary and the Cambridge Water case produced conflicting reasoning and decisions on first hearing and in the subsequent appeals. This gives rise to some interesting material for analysis in terms of the values (if any) which the judiciary bring to environmental issues in situations where they have considerable power to control the development of the law and the consequences which that may have for society.

The Cambridge Water case was heard by Mr Justice Ian Kennedy in the High Court in June 1991 (the judgment from the High Court case is reported at [1992] EnvLR 116). The case was based on three common law tortious actions: negligence, nuisance and the rule in Rylands v Fletcher. The main thrust of the Water Company's case was based upon the last of these actions and it is useful to explain and examine the history of the rule in Rylands v Fletcher (reported at [1861–73] All ER Rep 1). Rylands v Fletcher concerned the construction of a reservoir on the defendant's land; the defendant's contractors negligently failed to seal off shafts under the reservoir so that when it was filled, the water flooded the shafts which connected with a mine on the plaintiff's land which also became flooded. The flooding occurred without any fault on the part of the defendant, but nevertheless he was held liable by the Court of Exchequer Chamber (a forerunner of the Court of Appeal), Sir Colin Blackburn, delivering the judgment on the basis that anyone

> who for his own purposes brings on his lands and collects and keeps there anything likely to do mischief if it escapes, must keep it in at his peril, and, if he does not do so, is prima facie answerable for all the damage which is the natural consequence of its escape.

The rule, formulated as it was during the Industrial Revolution, is one manifestation of what appears to have been an attempt by the courts in the nineteenth century to protect the interests of landowners against potentially harmful activities carried out on neighbouring land. (For a review of nineteenth-century common law in relation to the environment see McLaren 1983. For an alternative social welfare analysis of Rylands v Fletcher see Simpson 1984.) The rule has become progressively emasculated by virtue of the subsequent judicial interpretation of an additional requirement to the rule, which was introduced by Lord Cairns in his speech (judgment) on the defendant's subsequent appeal to the House of Lords in 1868. This stipulates that in addition to the requirements specified by Mr Justice Blackburn, the defendant's user of their land must be 'non-natural'. In the context of the development of industrial activities in the nineteenth century, 'non-natural' user would give considerable scope to the judiciary to protect the environment. I do not pretend that there was some altruistic judicial ideal in this, but bearing in mind the social background of the nineteenth-century judiciary they may well have wished to protect the *status quo* of the landed classes to which they themselves belonged or originated from, and Simpson suggests a wider concern with social welfare. However, during the twentieth century a Porteresque 'anything goes' mentality seems to have prevailed and in our industrialised technological society it is extremely difficult to show that any user is 'non-natural'. This relaxed attitude is highlighted in the case of Read v Lyons (reported at [1947] AC 156), where the House of Lords took the unanimous view that the use of premises for the manufacture of explosives in an urban area was not a non-natural user. It had been wartime after all. Indeed, it would appear that the rule in Rylands v Fletcher has not been successfully argued in the English courts for the last 50 years.

The plaintiff Water Company in the Cambridge Water case may well have been regarded as indulging in some legal 'kite flying' in their use of Rylands v Fletcher, but this was the main basis of their claim. Almost inevitably within the formalist context of English law the trial judge felt obliged to comment on this (a pre-emptive strike?) by pointing out that it was infrequently used today, had been the subject of criticism by the Law Commission and had been limited by the significant decision of the House of Lords in Read v Lyons. The judge immediately turned to the issue of 'non-natural' user and, significantly, he refers to the village of Sawston as being 'industrial' which again pre-emptively goes to the heart of the plaintiffs' case on the issue of user. He also describes the defendants' business as 'useful' and emphasises the 350-year-old history of the trade in that particular village. There are no prizes for guessing what the judge will decide. The judge passes on to deal with the issues of 'non-natural' user arising from previous cases; non-lawyers may view with incredulity that this involves a review of cases from 1866 to 1969. But this would come as no surprise to an English lawyer and only emphasises the orthodoxy of the common law tradition, with all that entails in terms of limitation and closure. This review of the case law only serves to reinforce what the judge has pre-empted and the inevitable consequence is that he finds that the activities of the defendants are 'natural' and the plaintiffs' claim must therefore fail.

The judge does, however, seek some further justification outside of the facts of the case and the common law and he draws upon some policy issues. He refers again to Sawston as being an 'industrial village' and points to the benefits to the community of the employment which this industrial use affords. He goes no further than this and cynically states that he does 'not believe that I can enter upon an assessment of the point on a scale of desirability that the manufacture of wash leathers comes', and that:

> I reflect on the innumerable small works that one sees up and down the country with drums stored in their yards. I cannot imagine that all those drums contain milk and water, or some like innocuous substance. Inevitably that storage presents some hazard, but in a manufacturing and outside a primitive and pastoral society such hazards are part of the life of every citizen.

Apart from this utilitarian approach to the environment it is ironic that milk, which the judge regards as innocuous, is one of the most serious forms of water pollutant, causing twice as much oxygen depletion as raw sewage. A pollution incident involving milk would be of serious concern to the National Rivers Authority.

Despite the main concern in the High Court with the rule in Rylands v Fletcher, the plaintiffs also relied on the torts of nuisance and negligence. Nuisance has a long common law pedigree, and it was used in the nineteenth century to protect property interests. This had the 'knock-on' effect of environmental protection against industrial activity and pollution (McLaren 1983). The tort of negligence is of relatively recent origin and in its modern form originates in the well-known case of Donoghue v Stevenson (reported at [1932] AC 562). It is self-evident that these torts involve a civil 'wrong' which results in loss to the plaintiff. In order to 'control' the tort of negligence the courts require the damage or loss which is caused to be 'foreseeable'. This requirement of foreseeability has also crept into the tort of nuisance, although it is by no means clear that this was always a requirement and liability could be 'strict' (ie if the wrong was committed its harm has to be compensated). This turned out to be a contentious issue in the subsequent appeal to the Court of Appeal, and will be returned to later. It is sufficient to say here that in the High Court the judge took the contemporary view that the damage actually occasioned must be foreseeable. In his view the careless spillage of solvent might foreseeably cause someone to be affected by the fumes. His difficulty was in relating this to pollution. He states that:

> Pollution is in my judgement too wide a category. If I am right in drawing the inference that there were reasonably regular spillages of perhaps a few gallons, I am quite unpersuaded that the person responsible for an individual spillage ought reasonably to have foreseen any environmental damage.

He ends this part of this historical mental thought transfer exercise by devising a new test, ie:

> what would the reasonable supervisor overseeing the operation of the plant have foreseen to be the possible consequences of these repeated spillages over . . . perhaps ten years?

Note that the judge is stuck in the orthodoxy of the law. He still has to formulate his reasoning in the accepted manner that foreseeability is always to be tested by means of some form of reasonable person, be it the proverbial 'man on the Clapham omnibus' or 'ordinary reasonable supervisor'. There is no question that this may be an inappropriate approach to environmental issues. Orthodoxy, in the form of historical precedent and tried-and-trusted tests, sets the limits on judicial activity and creativity. Having set up the 'reasonable supervisor' test, the judge undertakes mental time travel into the mind of the supervisor at the defendants' premises and concludes that he would not have foreseen any environmental hazard. The whole judicial exercise seems vacuous in the light of the judge's statement that 'pollution . . . is too wide a category'. Again, in order to reinforce (justify?) what he says the judge then snipes at the medical evidence which lies behind the EC Directive on water quality which has precipitated this case. He states that even if his 'reasonable supervisor' foresaw environmental damage, they would not foresee the effect of organochlorines on water supplies. The judge draws on WHO papers from 1984 which show 'no evidence' of danger to human health from the consumption of water containing organochlorines, 'the concern being based on work with mice'. It is difficult to decide if the judge is truly blind to the possible dangers or just turning a convenient Nelsonian eye to them – whichever way it supports my underlying concern for judicial environmental myopia. Clearly the precautionary principle of environmentalism has no place here.

In essence the judge will not stray far from the orthodoxy of English law. Where he does so and considers wider environmental issues, it seems to be a cynical exercise. Sawston village is characterised as 'industrial', but employment follows from this. 'Pollution' is too wide for tort law to envisage. The effect of chemicals on laboratory animals seemingly has no consequences for human beings. The whole feeling of this judgment is that legal orthodoxy is safe, reasonable and capable of being tested; anything outside of this cannot be a basis for judicial reasoning. However, this does not deal with the issues of long-term environmental impairment which arise in this case. Indeed, the judge has no qualms in standing back from these issues when he says:

> There must be many areas of England and Wales where activities long ceased still have their impact on the environment, and where the perception of such impact depends on knowledge and standards which have been gained in more recent times. If it is right as a matter of public policy that those who where responsible for those activities, or their successors, should now be under a duty to undo that impact (or pay damages if a cure be impractical), that must be a matter for Parliament. The common law will not undertake such a retrospective enquiry.

This approach by the judge is mirrored in numerous 'hard decisions' in English law. Where a judge feels that he or she is treading on unsafe ground or is too cautious to develop the law, then the safe (accepted) retreat is to suggest that it is up to Parliament to reform the law. But even if the judge wished to stay within legal orthodoxy, he has overlooked the Single European Act 1986 (Article 130 Paragraph R) which contains the principle 'that the polluter should pay'. But perhaps this is an 'Act too far' for the judge.

It was inevitable that whichever way this case was decided, the losing side would appeal. Fifteen months later the parties' lawyers argued their respective cases in the Court of Appeal (reported at [1994] 1 ALL ER 62). The appellant Water Company pursued its appeal on the basis that if the trial judge had been correct in deciding that the storage of solvents at the defendants' premises was a 'natural' user of their land, then the rule in Rylands v Fletcher had been 'emasculated to the point of extinction'. It is clear that the bulk of the argument in the appeal had been exclusively on this sole point of the modern significance of the rule in Rylands v Fletcher. It is equally clear that the judges of the Court of Appeal did not really want to deal with this. It may be that they felt that the legal orthodoxy in the form of the way in which Rylands v Fletcher had been developed meant that they would have no alternative other than to agree with the trial judge's decision. Whatever the reason, the Court of Appeal were able to perform a legal conjuring trick by making the problem with Rylands v Fletcher disappear and producing a new prop with which they could work. Lord Justice Mann acknowledged:

> the questions raised . . . are fundamental ones in regard to the rule in
> Rylands v Fletcher, but they would not arise if the appellant could establish
> its claim under the heading of nuisance.

At a stroke the judge changes the basis of the case. It is difficult to comprehend the judge's reason for this. There is no hint in the main part of the judgment that the court had particular sympathy with one side or the other, and
there is little comment other than on substantive legal issues. The new magic
which the court is able to work is based on the case of Ballard v Tomlinson
(reported at [1885] 29 Ch D 114). This case was decided by the Court of
Appeal and technically the present Court of Appeal would be bound to follow it
in any subsequent similar case. In that case, the defendants having no use for a
well on their land used it as a convenient 'reservoir' for their sewage and had
run their sewerage pipes into it. Once the level had reached a certain height the
water table was such that sewage-contaminated water found its way into the
plaintiffs' well from which they abstracted water for their brewery. The present
Court of Appeal interpreted the decision of the 1885 Court of Appeal as being
that the polluter was liable irrespective of intent: 'it was sufficient that the
defendant's act caused the contamination'. This conveniently sidesteps the basis
on which the claims based on nuisance and negligence had failed in the High
Court. Indeed, the Court of Appeal chides both the trial judge and the lawyers
for overlooking the significance of this decision from 1886. On the basis that
this precedent was indistinguishable from that in the present case, the appeal
was upheld and damages in excess of £1 million were awarded against Eastern
Counties Leather.

It is interesting to note that yet again legal orthodoxy reigns supreme. By
means of a simple sleight of hand the outcome of the case is completely different, but it is all stage-managed within the confines of the law. At no stage in the
substantive judgment does the Court of Appeal need to consider any wider
issues, except that in the penultimate paragraph of the judgment the court
allows itself a little latitude. Here Lord Justice Mann mentions some policy
arguments advanced by the parties' respective QCs. The defendants had argued
that, as modern legislation both managed the resource and controlled the pollution of water, there was no need to have recourse to the common law. The
implication in this argument must be that if the modern legislation does not
cover the instant situation, then Parliament cannot have intended that it should
be, and the common law should keep out. The appellants countered that the
maxim 'the polluter pays' was enshrined in European Community law.
Surprisingly Lord Justice Mann makes no direct comment on these arguments,
but this is because (he says) there is no need to look beyond the law:

> where the law which is binding on this is clear, as we think it here is, then
> the court's decision cannot be affected by policy considerations.

The judgment concludes with the gratuitous and enigmatic comment:

> Others must consider whether Ballard v. Tomlinson accords with
> contemporary opinion. Some of them may say it does.

This comment is superfluous in the light of the formalistic nature of this judgment and it begs some speculation as to the motive behind it. Either the judge is indirectly responding to the defendants' policy arguments by an appeal to public opinion, or else it is a justification of his own adoption of a convenient 100-year-old case and his disregard of the foreseeability aspect. It is difficult to believe that an English judge may be expressing a personal view in this obtuse manner, but perhaps he has 20:20 vision when assisted by nineteenth-century common law spectacles.

The inevitable further appeal to the House of Lords by Eastern Counties Leather was heard in late 1993. The House of Lords, or more correctly the Appellate Committee of the House of Lords, sitting as the supreme appeal court of the UK, has the power to entrench the law because their decision on the point of law at issue will have the effect of binding all other UK courts for the future. Strictly speaking they do not bind themselves, but they will only over-rule their own decisions in exceptional circumstances. In effect, if they are per-ceived to have reached the wrong decision then only Parliament can legislate to change the law. The Committee comprised the usual number of five of the Law Lords, but the only substantive speech (judgment) was delivered by Lord Goff in December 1993 (reported at [1994] 1 ALL ER 62).

Lord Goff commences his speech by outlining the facts and history of the case. In an early paragraph he describes the solvents and emphasises their widespread and everyday usage both industrially and also in domestic situa-tions. He illustrates their domestic use with the example of 'Dab-it-off'. Clearly this emphasises the apparently innocuous nature of these solvents and this will be of great significance later in his speech. However, this does start to cloud the issues; there is no comparison with the few centilitres used to remove a persis-tent stain on one's clothing and the minimum estimate of 3200 litres spilt by Eastern Counties Leather. The maximum amount which they had spilt could not even be guessed at.

Having outlined the circumstances of the case, the style of a House of Lords decision is to examine the existing law closely. Lord Goff initially considers the relationship between nuisance and the rule in Rylands v Fletcher. He does this by introducing an article from an academic journal (Newark 1949) and then relying heavily on it. This article considers that the decision in Rylands v Fletcher has been misappreciated, and that the judges in that case did not con-sider they were creating a revolution in the law. All they were doing was apply-ing consistent principles of the tort of nuisance to the novel situation of an isolated escape of a harmful substance, whereas nuisances by their very nature would usually tend to be ongoing and not a 'one-off'. Again, a judicial conjur-ing trick produces a case involving Sir Colin Blackburn six years after his deci-sion in Rylands v Fletcher where he 'believed himself not to be creating new law, but to be stating existing law, on the basis of existing authority' (Ross v Feddon [1872] 26 Law Times at 966).

Having established this very close relationship it cleared the way for the issue of foreseeability to be examined as a requirement of both the tort of nuisance and also of the rule in Rylands v Fletcher. Dealing first with nuisance, Lord Goff is clearly influenced by the development of the tort of negligence. He states:

> But it by no means follows that the defendant should be held liable for damage of a type which he could not reasonably foresee; and the development of the law of negligence in the past sixty years points strongly towards a requirement that such foreseeability should be a prerequisite of liability in damages for nuisance, as it is in negligence.

He is also able to support a requirement of foreseeability from a 1966 decision of the Judicial Committee of the Privy Council (in effect the House of Lords by another name, sitting in its capacity of final appeal court for certain commonwealth countries) in the case of The Wagon Mound (No 2) (reported at [1947] AC 617). Privy Council decisions are not binding on the House of Lords, but are regarded as being very persuasive and Lord Goff regards this case as having:

> settled the law to the effect that foreseeabiliy of harm is indeed a prerequisite of the recovery of damages in private nuisance.

Lord Goff then deals with the tricky issue of foreseeability and the rule in Rylands v Fletcher. This is difficult to manage given the weight of judicial and academic belief that this was a tort of strict liability. He has already foreshadowed the outcome by setting up the misapprehension of the law which Professor Newark's article highlights. Lord Goff carefully examines the original judgment of Sir Colin Blackburn in Rylands v Fletcher and emphasises passages in the judgment which suggest that foreseeability was always an ingredient of the tort. It seems surprising that no one has spotted them before. He concludes that:

> The general tenor of his [Blackburn's] statement of principle is therefore that knowledge, or at least foreseeability of the risk, is a prerequisite of the recovery of damages under the principle.

Given that this is a major restatement of the tort, Lord Goff seeks further justification by considering the possible consequences of a rigorous application of strict liability. He recognises that this would have the practical benefit of fixing the cost of damage for hazardous activities on the overheads of the relevant businesses rather than the victim or society. The United States is cited as an example of such a trend, but his cursory examination of the relevant US legislation leads him to conclude that the dangerous activities referred to in the legislation would be obvious to any reasonable person who carried them on. A conclusion is unstated, but this conveniently supports his foreseeability arguments.

He gains further support from a 1970 Law Commission report. He emphasises that:

> the Law Commission expressed serious misgivings about the adoption of any test for the application of strict liability involving a general concept of especially dangerous or ultra-hazardous activity, having regard to the uncertainties and practical difficulties of its application. If the Law Commission is unwilling to consider statutory reform on this basis, it must follow that judges should if anything be even more reluctant to proceed down that path.

It seems ironic that a statutory body charged with keeping all law under review with a view to its systematic development and reform should be marshalled in support of such a negative stance.

The retreat of the trial judge to the safety of parliamentary reform is endorsed:

> Like the judge in the present case, I incline to the opinion that, as a general rule, it is more appropriate for strict liability in respect of operations of a high risk to be imposed by Parliament, than by the courts.

Lord Goff recognises the importance of environmental protection. He states that 'it is of crucial importance to the future of mankind', but ironically this is also the justification for the common law to do nothing. His view is that:

> public bodies, both national and international, are taking significant steps towards the establishment of legislation which will promote protection of the environment, and make the polluter pay for damage to the environment for which he is responsible . . . But it does not follow from these developments that a common law principle, such as the rule in Rylands v Fletcher, should be developed or rendered more strict to provide for liability in respect of such pollution. On the contrary, given that so much well-informed and carefully structured legislation is now being put in place for this purpose, there is less need for the courts to develop a common law principle to achieve the same end, and indeed it may well be undesirable that they should do so.

The difficulty with this approach is that the bulk of the legislation to which he refers is regulatory and not compensatory, and neither the UK nor the EU legislatures have dealt with the issues of historic pollution and its remediation. Lord Goff's arguments, from legal principle alone, fail to address the wider environmental issues which he himself has identified. The common law is adaptable to many new and difficult situations, as is demonstrated by the judicial conjuring tricks in the Cambridge Water case alone. It is to be regretted that the House of Lords has not had the courage of its stated environmental convictions to 'green' the common law.

The most depressing aspect of the House of Lords' involvement in this case is the response of the fifth member of the House of Lords, Lord Woolf. His speech is quoted in full:

> My Lords, I have had the advantage of reading in draft the speech prepared by my noble and learned friend Lord Goff of Chieveley. I agree with it and for the reasons he gives I too would allow the appeal.

CONCLUSION

It is ironic that the nineteenth-century judiciary were prepared to take a bold and far-sighted view of the role of the common law in terms of environmental (using the term very broadly) welfare and that this was in the age of *laissez-faire*. Their late twentieth-century successors in the 'caring 1990s' seem to lack any vision as to the potential role of the common law in situations of environmental impairment.

REFERENCES

Journal of Environmental Law (1992) **4**, pp1–4.

Law Commission (1970) *Civil Liability for Dangerous Things and Activities*, Law Commission, No 32.

McLaren, J. (1983) 'Nuisance Law and the Industrial Revolution – Some Lessons from Social History', *Oxford Journal of Legal Studies*, pp155–221.

Newark, F. (1949) 'The Boundaries of Nuisance', *Law Quarterly Review*, **65**, pp480–90.

Simpson, A. (1984) 'Legal Liability for Burst Reservoirs: the Historical Context of Rylands v Fletcher', *Journal of Legal Studies*, pp209–64.

11 Expert and Layperson Evaluation of Urban Environmental Quality: The 'Natural' versus the 'Built' Environment

MIRILIA BONNES AND MARINO BONAIUTO
Dipartimento di Psicologia dei Processi di Sviluppo e Socializzazione, Università degli Studi di Roma 'La Sapienza'

INTRODUCTION

Within environmental psychology, environmental assessment is a conceptual and methodological framework for describing and predicting relationships between the characteristics of places and the cognitive, affective, behavioural responses of people who use and evaluate those places. This field of enquiry is devoted both to a better understanding of environment–individual transactions and to improving environmental design and management (Craik and Feimer 1987). The main question within the field concerns the relation between technical and observational systems of environmental assessment. Technical assessment (which is often considered as 'objective') is grounded on measurement systems and indexes derived from specific disciplines and it does not require individual exposition. Observational assessment (which is typically considered as 'subjective') is grounded on descriptions and judgements expressed through ordinary language by several people about place attributes. Both of them aim to evaluate the same environment in which people live, but probably different values, ideals and goals about the environment underlie the expert's technical and layperson's observational assessments.

In fact, a clear-cut distinction between subjectivity and objectivity in environmental evaluation should not be taken for granted (Uzzell 1989): first of all, some degree of subjectivity is always involved in deciding which elements are significant and worthy of measurement; secondly, if person and environment are assumed to form an interdependent system in which they affect each other and are equally important, whichever one we are studying we should always consider it in relationship with the other.

Differences between societies (and within a society between different groups) in the representations of the environment and of environmental changes, as well

Values and the Environment: A Social Science Perspective, Edited by Yvonne Guerrier, Nicholas Alexander, Jonathan Chase and Martin O'Brien. © 1995 John Wiley & Sons Ltd.

as in human–environment relationships and in the management of environmental matters, reflect different cultural constructions and definitions of the environment (Douglas and Wildavsky 1982). In particular, the social-political diversity of the perception and evaluation of environmental events is evidence for the perspectival character of these social constructions (Graumann and Kruse 1990), which are functional to the different values and interests of the conflicting groups involved. Cultural, social and organisational groups can hold varied world views, values and opinions, which result in substantial differences in the groups' evaluations and responses to the impacts of local environmental changes (Nordenstam 1994). The discordances between different views about the environment often lie in the definitions of the environmental changes and concepts themselves. For example, a debate on the different proposals for a possible urban development focused mainly on the interpretation and explication of the meaning and the appropriate measures for the construct of 'urban quality' (Lalli and Thomas 1989). For the local government and the supporters of a modern centre, it was mainly a matter of the economic attractiveness of the town; for the supporters of the reconstruction of an old building, it was mainly a matter of symbolic meaning; while the opponents of any development intervention conceived of it in terms of the improvement of ecological and social resources. Actually, only a minority of the interviewed sample had opinions about the improvement of the local urban quality which were coherent with those on which the official development plans were predicated.

Even for apparently common-sense issues and shared concepts such as pollution, there can be disagreement about what should be intended as polluted or not, and when it becomes a risk for the environment and its population. Because there is no single accepted index of water, air or soil pollution, it is often hard to know just what is meant when pollution is being discussed; and this is partly due to the multidimensional character of many concepts referring to environmental features. Moreover, many of the consequences associated with certain environmental changes or phenomena are outside the traditional domains of economics or medical science: that is, outside easily quantifiable domains. The effects of environmental issues and problems often pertain to aesthetics and morality. Their conceptualisation and evaluation reflect the kind of world we want to live in and the kind of responsibility we feel about our environment. Human value systems are bound up with environmental matters and priorities. In this respect, environmental values should not be considered from a restricted viewpoint: the valuation of the environment in general, and of specific environmental aspects in particular (eg water, green areas, etc), cannot be reduced to a single dimension. Values cannot be stated simply in economic terms, that is, as a trade-off of money or goods which would be made for an environmental improvement. Recreational values (uses and activities afforded by the environment), contemplative and aesthetic values (knowledge and memories about the environment), functional values (well-being of people) and ecological values (well-being of other forms of life) cannot be fully reflected simply through a utility values approach (Coughlin 1976).

Therefore, people's perceptions and evaluations of environmental aspects and issues become central. The identification of environmental perception features is particularly interesting once we understand their links with physical measurements of the environment (Uzzell 1989). This would help in considering the value of environmental perception in changing or preserving the world we inhabit. For example, in the case of water pollution the comparison between 'objective' and 'subjective' measurements of the same place shows that there can be a fairly good relationship between objectively defined pollution and the subjective estimation of it, but the two evaluations do not overlap because users evaluate the water as less polluted with respect to biological measures (Moser 1984). Moreover, some objective measurements correlate more significantly with the perception of pollution than some other physical indexes. And generally, people's attitudes and preferences regarding a certain environment are far more complex and multidimensional; in the case of a stream or a lake, 'subjective' evaluations do not simply depend on the condition of the water (whether 'objective' or 'perceived'), but also on the characteristics of the surrounding land, including both natural and man-made elements, as well as their social connotations (Coughlin 1976).

Different models for the evaluation of environmental aspects may be considered along a *continuum* (Daniel and Vining 1983; Uzzell 1989): from models assuming a set of objective qualities 'out there', which may be readily identified by anyone and determine the quality of each place, to models increasingly based on the role of the perceivers as instrumental in the evaluation of each place.

This chapter – which is part of the UNESCO MAB-Rome Project (Bonnes 1991) – attempts to compare and integrate different analyses of the urban environment carried out by different disciplinary working groups, each applying a different model for the environmental assessment of a specific neighbourhood, or quartiere, of Rome. The study aims to compare, in particular, the experts' technical evaluation and the laypersons' perception of the quality of the urban environment, according to the results derived from the research of different disciplinary groups. The goal is to assess the degree of correspondence between technical measurements and public evaluations of the same residential urban environment.

For the laypersons' evaluations, we consider data derived from the environmental psychologists' working group concerning the environmental evaluation and residential satisfaction of the inhabitants. For the experts' evaluations, we consider data derived from different working groups: architects and urban planners for the spatial and functional features of the urban environment, for both the built and the 'green' aspects (Bagnasco and Arcidiacono 1991; Bizzi and Lapadula 1991); plant ecologists for the attributes of the green areas of the urban environment (Celesti Grapow and Fanelli 1991).

METHOD

SAMPLE

The research is focused on a neighbourhood of the city of Rome (quartiere Aurelio), built mainly during the 1950s and 1960s, located in the north west part of the city and situated midway between the central area (historic centre) and the outer peripheral area. The neighbourhood is rather heterogeneous so far as its spatio-physical and population distribution is concerned. Following the analysis carried out by the town-planning work group, it has been subdivided into six different 'zones', each of which has a greater degree of spatial and town-planning homogeneity (Bagnasco and Arcidiacono 1991).

The interview sample consisted of 461 inhabitants whose homes were distributed over these six zones. First, 30 inhabitants were interviewed; subsequently, a structured questionnaire was submitted to 461 inhabitants. The quota sample was distributed within the six areas of the Aurelio neighbourhood according to gender, age, socioeconomic level, temporal experience and time spent daily in the neighbourhood (Bonnes and Bonaiuto 1991).

LAYPERSONS' EVALUATIONS

The environmental psychologists' working group developed different measures concerning the environmental evaluation and residential satisfaction of the inhabitants. They refer to the different aspects of the urban residential environment resulting from a factor analysis carried out on the answers to a 47-item questionnaire (Bonnes and Bonaiuto 1991). Three factors were extracted:

1. *Factor 1: 'Spatio-social density'* This refers to perceived density (closedness/openness) qualities. This applied both to the spatial features of the environment (in architectural town-planning terms: non-density of buildings, balance between built-up and open areas, volume and height of buildings, width of streets, possibility of exploiting parks and gardens) and to the social features of the neighbourhood (non-crowded conditions, inhabitants' sociability).

2. *Factor 2: 'Socio-spatial security/friendliness'* This dimension mainly involves evaluation of the friendliness/security of the social (heterogeneous population of different backgrounds, juvenile deviance) and spatial (lack of meeting places, isolation and distance from the centre) aspects of the environment.

3. *Factor 3: 'Socio-functional facilities'* This refers to the evaluation of functional facilities (specialized functional structures such as shops, entertainment, sports, etc) and of social responsiveness (willingness of the inhabitants to form interpersonal relations) in the neighbourhood itself.

Subsequently, another factor analysis was carried out within the 14 items composing the first evaluative factor. This further factor analysis was motivated mainly by the multidisciplinary frame of the research, in order to have more specific laypersons' evaluative dimensions to be directly comparable with some of the experts' environmental quality evaluations. Three sub-factors were extracted; they distinguish more specific modalities according to which perception of spatio-social density was structured in the laypersons:

1. *Sub-factor 1.1: 'Physico-static spatial density (closedness)'* This density can be defined as being of the physico-static type with reference to the evaluation of spatial architectural town-planning density.
2. *Sub-factor 1.2: 'Human-dynamic spatial density (closedness)'* This density can be defined in relation to the evaluation of crowding, including traffic.
3. *Sub-factor 1.3: 'Spatio/green-social facilities (openness)'* This refers both to spatial (ie availability of green spaces) and social (ie cordiality of people) receptiveness or environmental openness/closedness.

EXPERTS' EVALUATIONS

The other disciplinary teams developed different measures of environmental quality for the six residential zones. Here, we were able to utilise seven of these measures of objective environmental quality; they refer to specific aspects of the environment.

As far as the 'built environment' is concerned, the urban-planners group (Bagnasco and Arcidiacono 1991) provided:

1. *Index of territorial building volume* (m^3/m^2).
2. *Index of tertiarisation* (number of non-residential services and activities per 100 inhabitants).

Moreover, the urban-planners group (Bizzi and Lapadula 1991) supplied a measure of *degree of functional quality and centrality* of the census sections composing the six residence areas (number of rare and common services in each section). For each of the six areas, we derived an *index of functional centrality* as a ratio of functional quality of the area on area's extension:

$$[(\text{census section} * \text{rank}^2)] \div (\text{Ha/number of census sections})$$

In addition, a measure for the *degree of human density* was deduced from the 1981 census data, with respect to the same areas. For each of the six areas, we derived an *index of population density* as a ratio of the number of inhabitants per hectare (Ha).

As far as the *'natural environment'* is concerned, the plant ecologists group (Celesti Grapow and Fanelli 1991) supplied: measures of *degree of green areas quality* (flora and vegetation quality) for the same areas. According to these measures, the areas were ranked along a gradient of decreasing naturalness of flora and vegetation.

Moreover, the urban-planners group (Bagnasco and Arcidiacono 1991) provided a measure of *endowment of green areas* (percentage of public and private green areas in relation to total area); and a measure of *accessibility to urban public green areas* (distance from nearest urban public park).

In these seven cases the experts' technical information gave us an empirical ordering for the six zones which was sufficiently clear and univocal. Thus, each index gives the possibility of ranking the six residence areas according to different criteria, from the higher to the lower area along each of the qualitative hierarchies.

STATISTICAL ANALYSES

Factor scores for each of the main evaluation factors and sub-factors were subjected to one-way analyses of variance using the zone of residence inside the neighbourhood as the independent variable. The resulting average satisfaction scores for each area gave us the possibility of ranking the six areas along every evaluative dimension. Thus, the six zones were ranked according to two independent criteria: a 'subjective' criterion (ie according to inhabitants' residential satisfaction as expressed by a specific factor score) and an 'objective' criterion (ie according to environmental quality as expressed by the correspondent experts' technical evaluation). Then we compared laypersons' and experts' ranking (using Spearman r_s) in order to have a measure of their correlation for each aspect.

RESULTS

A significant effect for residence area emerged constantly on all the residential satisfaction scores. For each of the specific evaluative dimensions, the average scores are different between the inhabitants of the six areas: 'spatio-social density' ($F(5421) = 4.38$, $p<.001$); 'physico-static spatial density' ($F(5421) = 3.81$, $p<.01$); 'human-dynamic spatial density' ($F(5421) = 4.27$, $p<.001$); 'spatio/green-social facilities' ($F(5421) = 3.19$, $p<.01$); 'socio-spatial security/friendliness' ($F(5421) = 8.86$, $p<.001$); 'socio-functional facilities' ($F(5421) = 3.81$, $p<.01$). Mean values are reported in the fourth column of Tables 11.1a–d and 11.2b–c, and in the third column of Table 11.2a.

Tables 11.1 and 11.2 show the rank order of the six areas according to the experts' evaluations of the residential environment and the relative laypersons' perceptions of residential satisfaction. Both for built and natural features of the environment, Spearman r_s does not reach significance, although in certain cases it shows some relation between the experts' index of environmental quality and laypersons' residential satisfaction. Particularly in the case of man-made or built features of the environment such as physical spatial density (Table 11.1a), human spatial density (Table 11.1b), and quality of functional centrality (Table 11.1d), there is an overlap between experts' evaluation and users' perception in the lower ranks. This indicates that the lowest levels of environmental quality correspond to the lowest levels of residential satisfaction; with the index of functional centrality corresponding more closely than the index of tertiarisation to residents' evaluation of socio-functional facilities. On the contrary, no overlap but some negative correlation was found between experts' evaluation of the natural features of the environment and laypersons' satisfaction for these aspects (Table 11.2a–c).

Table 11.1   Built features of the environment: rank order of the six areas according to the experts' evaluation of urban environmental quality and the laypersons' perception of residential satisfaction

11.1a   Quality of physico-spatial density (index of territorial building volume) and perceived satisfaction (sub-factor 1.1: physico-static spatial density)

Residence area	Territorial building volume (m^3/m^2)	Rank	Degree of satisfaction for physico-static density	Rank	$r_s = .543$ n.s
1	1.53	6	0.64	3	
2	4.50	1	−10.04	2	
3	2.02	4	1.69	4	
4	2.93	2	−43.54	1	
5	1.87	5	29.22	6	
6	2.29	3	11.44	5	

11.1b   Quality of human-spatial density (index of population density) and perceived satisfaction (sub-factor 1.2: human-dynamic spatial density)

Residence area	Population density (inhab/ha)	Rank	Degree of satisfaction for human-dynamic density	Rank	$r_s = .600$ n.s
1	69	6	4.52	4	
2	311	1	−23.50	1	
3	138	4	−1.33	2	
4	164	3	15.25	5	
5	82	5	38.55	6	
6	170	2	4.07	3	

11.1c Functional quality (index of tertiarisation) and perceived satisfaction (factor 3: socio-functional facilities)

Residence area	Index of tertiarisation (no of services per 100 inhab)	Rank	Degree of satisfaction for services	Rank	$r_s = .090$ n.s
1	1.09	1	−38.77	1	
2	3.00	4	12.66	5	
3	2.51	2	31.45	6	
4	3.17	6	−18.36	2	
5	3.02	5	−9.28	3	
6	2.98	3	12.38	4	

11.1d Functional quality (index of functional centrality) and perceived satisfaction (factor 3: socio-functional facilities)

Residence area	Index of functional centrality (funct qual/extens)	Rank	Degree of satisfaction for services	Rank	$r_s = .771$ n.s
1	2.64	1	−38.77	1	
2	35.84	6	12.66	5	
3	19.73	4	31.45	6	
4	18.69	3	−18.36	2	
5	15.41	2	−9.28	3	
6	35.47	5	12.38	4	

Table 11.2 Natural-features of the environment: rank order of the six areas according to the experts' evaluation of urban environmental quality and the laypersons' perception of residential satisfaction

Table 11.2a Quality of green areas (degree of flora and vegetation quality) and perceived satisfaction (sub-factor 1.3: spatio/green-social facilities)

Residence area	Rank of flora and vegetation quality	Degree of satisfaction for green areas	Rank	$r_s = -.429$ n.s
1	6	−32.41	1	
2	1	7.22	3	
3	2	26.47	6	
4	4	−26.45	2	
5	5	15.49	5	
6	3	8.30	4	

11.2b Quality of green areas (endowment of green areas) and perceived satisfaction (sub-factor 1.3: spatio/green-social facilities)

Residence area	Endowment of green areas (% of total area)	Rank	Degree of satisfaction for green areas	Rank	$r_s = -.429$ n.s
1	>30.0	6	−32.41	1	
2	0.0	1	7.22	3	
3	13.0	2	26.47	6	
4	27.3	4	−26.45	2	
5	30.0	5	15.49	5	
6	16.5	3	8.30	4	

11.2c Quality of green areas (accessibility to urban public green areas) and perceived satisfaction (sub-factor 1.3: spatio/green-social facilities)

Residence area	Accessibility to urban public green areas (average distance in m)	Rank	Degree of satisfaction for green areas	Rank	$r_s = -.660$ n.s
1	102	6	−32.41	1	
2	417	3	7.22	3	
3	392	4	26.47	6	
4	325	5	−26.45	2	
5	531	1	15.49	5	
6	431	2	8.30	4	

DISCUSSION

Correspondences between the laypersons' environmental perceptions and the experts' evaluations of the same environment show in general a weak correlation between these two kinds of measures.

On the one hand, the inhabitants' residential satisfaction seems very sensitive to the environmental features of the residential place: significant differences have been found in residents' satisfaction according to the six different zones of residence. The subdivision of the neighbourhood indicated by the town-planning working group on the basis of a global evaluation given by them about the environmental/architectural features of the different zones found a strong correspondence in all the evaluative dimensions of the residents.

On the other hand, the attempt to find more specific correspondence between the residents' evaluations of particular environmental features and the objective conditions defined by the experts' evaluations of the same features did not produce significant results (Tables 11.1 and 11.2).

As far as the evaluation of the built environment is concerned, the highest correlations were found between: the index of functional centrality of the area and satisfaction for socio-functional facilities; the index of population density and satisfaction for human-dynamic spatial density; and the index of territorial building volume and satisfaction for physico-static spatial density. Moreover, in these cases the correspondence between experts' evaluation and laypersons' perception is particularly evident at the lowest rank orders (1st or 2nd), that is, in the case of poorer environmental quality. The correspondence between these two kinds of evaluation is less clear in the case of medium or high quality ranks, that is, where the environmental quality should be medium or high.

As far as the evaluation of the natural environment is concerned, referring to the quality of the green areas of the neighbourhood, correlations between the two kinds of evaluation tend to be weaker with respect to the previous ones; moreover they are always negative. Natural features, such as the green and plant features of the urban environment, seem to have a very different evaluative frame of reference for the users with respect to the experts (both natural scientists and urban planners). Other studies carried out on the same neighbourhood (Ardone and Bonnes 1991; Bonnes et al 1990a) have shown the multimodal character of the residents' relationship with the urban green areas. The physical, behavioural, social and symbolic aspects are strictly linked in the psychological construction of these urban features. The green aspects of the city seem to figure as specific urban 'sub-places' inside the place-neighbourhood (Bonnes et al 1990b), in which both the physical and the social aspects of the city inhabitants' relationships are integrated by users' activities and representations. On this basis, it is possible to understand the lack of correspondence between the kind of indexes of green area quality as provided by experts, and the residential satisfaction for the spatio/green-social facilities which involves a multimodal character in relation to the environment and in its use. In this case, the experts' and laypersons' values about what characterises a 'good' natural urban environment seem to vary significantly, given the different contribution of the functional aspects implied by their different evaluation models. If people's environmental uses and activities have a central role for their environmental perceptions and evaluations, then environmental assessment in general should devote more attention to studying what a place affords to the people who use it, that is, to highlighting the action possibilities of the environment and how these govern the public's perceptions and evaluations of that environment (Uzzell 1989).

On the whole, the results again point out the very problematic nature of the relations between technical (experts') and observational (laypersons') measures of environmental evaluation, and in particular their correspondence for very specific features of the environment. This is coherent with observations which showed other kinds of divergences resulting from the clash of different points of view on some environmental issues. In fact, issues of political and social values become fundamental even between experts, in cases of dispute or disagreement: for example Frankena (1983), analysing a renewable energy siting dispute, found that when different experts disagree they focus less on factual information, shifting the emphasis from technical issues to explicit value issues.

A general lack of correspondence is also usually reported by studies which compared the perceptual and cognitive evaluations of the public with the subjective evaluations of either architects (Devlin 1990; Groat 1982) or urban planners (Hubbard 1994). Even subjective interpretations and assessments by specific professional groups of urban environmental features through visual categorisation tasks, thus avoiding language mediation, did not show results congruent with those of the wider population. Therefore public–expert disagreements, and also expert–expert or politician–expert disagreements, can be conceived of as conflicts between different judgements often based on both terminological differences and legitimate conflicts of interests and values (Fischhoff et al 1987). Urban decision processes should take into account these different perspectives and try to integrate them, with the support of environmental perception and evaluation studies.

Furthermore, our results show the need for a closer collaboration between the different disciplines working on the social and physical aspects of the environment, in particular between natural and social sciences which often have different underlying value systems. Only through such collaboration could a complete descriptive and evaluative analysis of residential urban environment characteristics be accomplished.

ACKNOWLEDGEMENTS

We thank Prof Anna Paola Ercolani for her help and suggestions in data analysis.

REFERENCES

Ardone, R.G. and Bonnes, M. (1991) 'The Urban Green Spaces in the Psychological Construction of the Residential Place', in M. Bonnes (ed) *Urban Ecology Applied to the City of Rome*, MAB-UNESCO Project 11, Progress Report no 4, MAB Italia, Roma, pp149–73.

Bagnasco, C. and Arcidiacono, I. (1991) 'Subdivision of the Aurelio District into Homogeneous Zones and their Characteristics', in M. Bonnes (ed) *Urban Ecology Applied to the City of Rome*, MAB-UNESCO Project 11, Progress Report no 4, MAB Italia, Roma, pp67–89.

Bizzi, G. and Lapadula, B. (1991) 'Hierarchy of Activities for the Analysis of Urban Areas', in M. Bonnes (ed) *Urban Ecology Applied to the City of Rome*, MAB-UNESCO Project 11, Progress Report no 4, MAB Italia, Roma, pp45–66.

Bonnes, M. (ed) (1991) *Urban Ecology Applied to the City of Rome*, MAB-UNESCO Project 11, Progress Report no 4, MAB Italia, Roma.

Bonnes, M. and Bonaiuto, M. (1991) '"Subjective" and "Objective" Evaluations of the Quality of Urban Environment: Some Comparative Results', in M. Bonnes (ed) *Urban Ecology Applied to the City of Rome*, MAB-UNESCO Project 11, Progress Report no 4, MAB Italia, Roma, pp175–91.

Bonnes, M., De Rosa, A., Ardone, R. and Bagnasco, C. (1990a) 'Perceived Quality of Residential Environment and Urban Green Areas', *Blraum-Blauguetia*, **3**, pp54–62.
Bonnes, M., Mannetti, L., Secchiaroli, G. and Tanucci, G. (1990b) 'The City as a Multi-place System: from Intra-place to Inter-place Analysis of People–urban Environment Transactions', *Journal of Environmental Psychology*, **10**, pp37–65.
Celesti Grapow, L. and Fanelli, G. (1991) 'Ecological Study on Flora and Vegetation of the Aurelio Area in Rome', in M. Bonnes (ed) *Urban Ecology Applied to the City of Rome*, MAB-UNESCO Project 11, Progress Report no 4, MAB Italia, Roma, pp103–24.
Coughlin, R. (1976) 'The Perception and Valuation of Water Quality: a Review of Research Method and Findings, in K. Craik and E. Zube (eds) *Perceiving Environmental Quality: Research and Applications*, Plenum, New York, pp205–27.
Craik, K. and Feimer, N. (1987) 'Environmental Assessment', in D. Stokols and I. Altman (eds) *Handbook of Environmental Psychology, Vol 1*, Wiley, New York, pp891–918.
Daniel, T. and Vining, J. (1983) 'Methodological Issues in the Assessment of Landscape Quality', in I. Altman and J. Wohlwill (eds) *Human Behavior and Environment: Vol 6*, Plenum, New York, pp39–84.
Devlin, K. (1990) 'An Examination of Architectural Interpretation: Architects versus Non-architects', *Journal of Architectural and Planning Research*, **7**, 3, pp235–44.
Douglas, M. and Wildavsky, A. (1982) *Risk and Culture. An Essay on the Selection of Technological and Environmental Dangers*, University of California Press, Berkeley, CA.
Fischhoff, B., Svenson, O. and Slovic, P. (1987) 'Active Responses to Environmental Hazards: Perceptions and Decision Making', in D. Stokols and I. Altman (eds) *Handbook of Environmental Psychology, Vol 2*, Wiley, New York, pp1089–133.
Frankena, F. (1983) 'Facts, Values, and Technical Expertise in a Renewable Energy Siting Dispute', *Journal of Economic Psychology*, **4**, pp131–47.
Graumann, C. and Kruse, L. (1990) 'The Environment: Social Construction and Psychological Problems', in H. Himmelweit and G. Gaskell (eds) *Societal Psychology*, Sage, London, pp212–29.
Groat, L. (1982) 'Meaning in Post-modern Architecture: an Examination Using the Multiple Sorting Task', *Journal of Environmental Psychology*, **2**, pp3–22.
Hubbard, P. (1994) 'Diverging Evaluations of the Built Environment: Planners versus the Public', in S. Neary, M. Symes and F. Brown (eds) *The Urban Experience: a People–Environment Perspective*, E & FN Spon, London, pp125–33.
Lalli, M. and Thomas, C. (1989) 'Public Opinion and Decision Making in the Community: Evaluation of Residents' Attitudes towards Town Planning Measures', Urban Studies, **26**, pp435–47.
Moser, G. (1984) 'Water Quality Perception, a Dynamic Evaluation', *Journal of Environmental Psychology*, **4**, pp201–10.
Nordenstam, B. J. (1994) 'When Communities say NIMBY to their LULUs: Factors Influencing Environmental and Social Impact Perception', 14th Annual Meeting of the International Association for Impact Assessment, Quebec, Canada, June 14–18.

Uzzell, D. (1989) 'People, Nature and Landscape: an Environmental Psychological Perspective', Report for the Landscape Research Group, University of Surrey, Guildford.

Section IV

CHANGING
ENVIRONMENTAL VALUES

Introduction

MARTIN O'BRIEN
Department of Sociology, University of Surrey

As earlier chapters of this book have made abundantly clear, 'environmental values' can be understood in many different senses: as economic values placed on specific resources or environmental goods; as political values attached to particular locations or lifestyles; as social values circulating within and between different human communities which establish status categories or rights of environmental access; as personal values interconnected with wider frameworks of belief and moral commitment; as spiritual values underpinning codes of cultural conduct; and many others. This diversity presents problems for activists who seek to change 'values' as a strategy for protecting or repairing the environment. The problems stem from the intersection of economic, political, psychological, social and cultural processes in the creation and maintenance of those values. Values do not arise in a vacuum and value change does not occur unless it has some benefit or utility for people. Values are developed, adhered to, rejected, dismissed and reformulated for particular reasons, under particular circumstances and at particular times. If this were not the case there would be little point in attempting to change values: values are addressed precisely because they offer the possibility of effecting change in social, economic and political behaviours, logics and benefits. The adoption of one set of value criteria over another implies a much wider reorientation in people's lives: in their patterns of work and leisure, production and consumption, public and private behaviours and in political commitments and opportunities. Furthermore, the values apparently held by any individual are themselves unstable and shifting rather than fixed and reliable: they embody contradictions, inconsistencies and wide variations in different circumstances. This is not because people are fickle or ill informed. It is because different values are useful to people in different ways, at different times, in different situations. People draw upon values as resources for achieving the goals they have in their public and private lives and seek actively to construct value frameworks which support those goals.

In industrialised, multiethnic, socially segmented societies such as the United Kingdom, environmental values are forged in and channelled through a bewildering diversity of institutional, communal and interpersonal networks. Some dimensions of people's value frameworks are conditioned by their experiences of technological and economic organisation: the necessity to work for a

Values and the Environment: A Social Science Perspective, Edited by Yvonne Guerrier, Nicholas Alexander, Jonathan Chase and Martin O'Brien. © 1995 John Wiley & Sons Ltd.

wage in order to survive, the mediation by machinery of people's relationships to nature (and to each other), and the accompanying distance which these forces establish between individuals and ecosystems encourage a rationalising and objectifying orientation to the problem of environmental commons. Other dimensions are conditioned by the social and political divisions which pervade modern life: divisions in public and private experience, distinctions of wealth, status and access to resources and the distance which these forces establish between groups of people within and between societies encourage contradictory and conflicting commitments, priorities and alliances over the consequences of environmental change. Still other dimensions are conditioned by the symbolic and aesthetic significance which features of the environment embody in different cultural traditions: the different ways in which the countryside is revered in British conservatism contrasts sharply with the spiritual values placed on forest and land by the Indian Chipko movement (see Shiva 1989); the pseudo-religious veneration of gain and wealth in capitalist cultures is contested by innumerable communal and tribal values emphasising the rejection of (so-called) materialist world views in favour of harmonious and sensuous relationships between humans and environments.

To acknowledge this array of influences on and contexts of environmental value formation is to recognise that changing environmental values is a task of monumental proportions. Whereas common sense may urge a focus on personal value change as a relatively simple and effective strategy for dealing with problems of overconsumption, excessive use of the motor car, the indiscriminate waste of resources in industrial production, and so on, even limited reflection on the characteristics of value frameworks indicates that such common-sense beliefs do not really address the social complexity in which the former are irredeemably embroiled. Changing values requires action at the institutional, personal, political and economic levels simultaneously. A strategy for changing personal values needs to be accompanied by the development of social contexts in which those changed values can be meaningful to and useful for people. The arts of persuasion may be powerful, but they are not, on their own, powerful enough.

Some of these issues arise, albeit in different forms, in the chapters in this final section of the book, which is concerned with one location where inroads into environmental value formation can be made – in education. The three chapters address different elements of this problem. Uzzell et al report on research into the effectiveness of 'hands-on' environmental learning techniques. They raise a number of key questions about environmental attitudes and ethics and call for a critical reappraisal of the assumptions underlying the philosophy and practice of environmental education. This critical reappraisal is of some urgency in the UK since environmental education is now a core, cross-curricular element of the National Curriculum and is targeted precisely at the formation of beliefs about and attitudes towards the environment. However, as the authors point out, the contemporary commitment to hands-on, experiential education on environmental matters may be misplaced. Their research shows that the results of such educational programmes are ambivalent and that this ambivalence is generated, in important ways, by the contradictory push and pull of surrounding political, economic and cultural forces. In particular, they suggest that current models of environmental education may reinforce divisions between scientific representations of events and social representations of events. The consequence of these divisions is to encourage a separation between the worlds of science and the everyday worlds which people inhabit. In the case of environmental science this poses the danger that the change sought by the education programme may be exactly the opposite of the change achieved by its implementation.

While Uzzell et al examine the characteristics of secondary education, the following two chapters contribute to the debate on environmental higher education. Dibble makes a case for the development of a specialist, compulsory course aimed at generating environmental responsibility. In the tradition of Friere (1973) and Lipman (1991), Dibble proposes that environmental responsibility can only emerge from the interests and knowledge held by people themselves and, by implication, cannot be imposed from outside the contexts in which such knowledge and interests are developed and formulated. His model course in environmental ethics is based on Lipman's ideal of the community of inquiry in which the process of formulating questions and sharing strategies for finding answers to them is more important than the dissemination of expert knowledge about the science of environmental change. Such a model raises serious resource dilemmas for the ways in which higher education is currently conducted, and it will be interesting to see how institutions respond to the challenge that Dibble issues.

Dibble makes a general call for new educational strategies for environmental education, and the final chapter by Haigh provides an insight into how learners may be encouraged not only to understand competing value frameworks but also to apply those value frameworks to the environments in which they live. The approach outlined by Haigh contrasts markedly with the approach criticised by Uzzell et al and involves hands-on experience not primarily with the physical constituents of natural environments but with their cultural, aesthetic and social meanings. In this sense, Haigh provides a practicable way forward for environmental educators seeking to make an impact not only on the ways in which learners understand the environment, but also how they understand alternative cultural world views and representations and alternative commitments to natural and social phenomena. Opening minds to alternative explanations is precisely one of the functions of a higher education system.

Haigh's chapter on environmental learning provides a well-crafted model for the use of critical techniques in social science education in exploring some of the many different patterns of values which persist within and between cultures. Education about environmental values cannot resolve all of the dilemmas to which I pointed in my introductory remarks. It can, however, provide a stimulus for change in certain limited fields of environmental action and awareness. If the call to change values is to become more than a worn-out cliché, new strategies in the field of education must be accompanied by simultaneous change in related key organisations and institutions. The processes of learning are as important as the outcomes of learning, and currently these processes are embedded in dominant ideals of efficiency, effectiveness and economy which gloss over many of the problematic learning experiences which today pass for 'environmental education'. Without genuine and committed attention to such processes, and to the ideals and contexts through which they are sustained, the developments in education for change outlined here will encounter political obstacles and barriers which will render them less effective than they, and we, deserve.

REFERENCES

Friere, P. (1973) *Education for Critical Consciousness*, Seabury Press, New York.
Lipman, M. (1991) *Thinking in Education*, Cambridge University Press, Cambridge.
Shiva, V. (1989) *Staying Alive: Women, Ecology and Development*, Zed Books, London.

12 Questioning Values in Environmental Education

DAVID L. UZZELL, ADAM RUTLAND AND DAVID WHISTANCE
Department of Psychology, University of Surrey

INTRODUCTION

Two assumptions more than any other have dominated thinking in environmental education. First, there is the widespread assumption that an experiential approach to environmental education is the most powerful and effective way of enhancing understanding of environmental issues and processes. Secondly, it is assumed that children should be the key audience for environmental education and encounters because they are seen as tomorrow's opinion leaders and stewards of the earth (Sutherland and Ham 1992). Environmental education has been set the task, therefore, of changing children's values, attitudes and behaviours through environmental encounters and the communication of 'hard' scientific facts.

International and intergovernmental support and advocacy for environmental education have a long pedigree, including: the IUCN conference on 'Environmental Education and the School Curriculum' (IUCN 1970), the UNEP/UNESCO International Environmental Education Programme (1975), the UNESCO Tbilisi Conference on Environmental Education (1977), the UNESCO 'Tbilisi Plus Ten' Conference on Environmental Education (1987), the World Commission on Environment and Development (1987), the Council of the European Community (1988) and most recently the Earth Summit Conference in Rio de Janeiro (UNCED 1992). The importance of environmental education has also been recognised within individual states in the European Union. In September 1990, the UK government White Paper, *This Common Inheritance: Britain's Environmental Strategy* (Department of the Environment 1990) made the case that education has a critical part to play in raising the public's awareness about local and global environmental problems. It was also stated that the 'educational system must play an important part in promoting environmental awareness, understanding and competencies'.

How are such objectives to be achieved? It was the Plowden Report (Department of Education and Science 1967) which first highlighted the importance of using the environment as a resource for learning within schools. Local

Values and the Environment: A Social Science Perspective, Edited by Yvonne Guerrier, Nicholas Alexander, Jonathan Chase and Martin O'Brien. © 1995 John Wiley & Sons Ltd.

education authorities and individual schools have had the responsibility of deciding when and how environmental education appeared in the curriculum. Under the 1988 Education Reform Act (Department of Education and Science 1988a) the UK government specified what should be taught in schools by establishing a National Curriculum.

Environmental education is one of the officially recognised and documented cross-curricular themes of the National Curriculum (the other areas are health education, education for citizenship, careers education and guidance, and economic and industrial understanding). These cross-curricular themes are considered central elements of the whole curriculum, with consistent aims and a clear notion of how children's education in these areas should progress. Environmental education, as a cross-curricular theme, can occur in or develop out of various areas of the curriculum such as science, geography and history. What distinguishes these cross-curricular themes from the rest of the curriculum is that they promote the discussion of values and beliefs. In addition, they rely on practical experiential learning and decision making. As previously, schools have been free to interpret the government guidelines on how environmental education is best taught within the whole curriculum: 'in primary schools they can be adapted either to the theme or topic approach . . . or to more formal subject teaching' (National Curriculum Council 1990a).

THE GOAL OF ENVIRONMENTAL EDUCATION IN ENGLISH AND WELSH SCHOOLS

There are three principles which guide environmental education in England and Wales. These were first formulated by the Schools Council's Project Environment in 1974. They are characterised by the notions of education *about*, *for* and *in* the environment:

1. Education *about* the environment aims to develop children's knowledge and understanding about values and attitudes.
2. Education *for* the environment puts the emphasis on children developing an informed concern for the environment. The ultimate aim is for all children to develop a personal environmental ethic which will lead them into actions to benefit the physical environment.
3. Education *in* or *through* the environment views the environment as a useful resource for learning. It allows children to develop knowledge and understanding plus the skills of investigation and communication.

Embedded within these three principles are the specific aims of the learning process, the development of knowledge, skills, attitudes and behaviour. These are referred to and articulated in the documents of both statutory and advisory organisations which attempt to define the aims and content of environmental education (Council for Environmental Education 1987; Department of Education and Science 1988a; Department of Education and Science 1988b;

National Curriculum Council, 1990b). For example:

Knowledge and Skills

i). To develop a coherent body of knowledge about the environment, both built and rural, sufficient to recognise actual and potential problems,

ii). To be able to gather information from and about the environment independently or as part of co-operative activity,

iii). To be able to consider different opinions related to environmental issues and to arrive at a balanced judgement,

iv). To appreciate the ways in which environmental issues are interrelated so that one factor affects others,

v). To be able to evaluate information about the environment from different sources and to try to resolve environmental problems,

vi). To understand and to know how to use the mechanisms available in society for bringing about environmental change.

Attitudes and Behaviour

i). To develop an appreciation of the environment and critical awareness of the natural and built environment,

ii). To develop an attitude of concern for environmental matters and a wish to improve environmental understanding,

iii). To be critical of one's own environmental attitudes and to take steps to change one's own behaviour and actions,

iv). To have a desire to participate in initiatives to care for or improve the environment,

v). To wish to participate in environmental decision making and to make opinions known publicly.

(Council for Environmental Education 1987)

As Palmer and Neal (1994) argue, a broad consensus has emerged concerning the principles and objectives of environmental education, and central to all are notions of raising awareness and understanding, developing values, attitudes and skills and, in some cases, engaging in behaviours which are pro-environmental.

CRITIQUE OF CURRENT ENVIRONMENTAL EDUCATION PRACTICE AND RESEARCH

It could be argued that environmental education has for too long been hampered by a well-meaning but ultimately meaningless rhetoric that should be subject to critical appraisal. Even as late as 1990, views such as that expressed in the Gilbert Report – 'The ultimate aims of environmental education are the creation of responsible attitudes and the development of an environmental ethic' (Scottish Education Department 1974) – were still being endorsed (Hale 1990). What are *responsible* attitudes? Responsible in whose eyes? What is an environmental ethic? A critical appraisal is needed of the values and assumptions which currently inform the philosophy and practice of environmental education. This will only emerge from a theoretically driven programme of research (ESRC 1992).

Environmental education sometimes has a questionable ethical basis where behaviour changes are made the primary goal, while the acquisition of knowledge and people's own decisions are pushed into the background. This often stems from a teaching and learning model which is top down and centre to periphery. It is very easy when advocating a cause which is so 'obviously' correct to prescribe with dubious certainties and ready-made solutions. The actual content of the environmental message is not as important as how the environmental problem is presented to children. *How* the issue is addressed indeed could be more crucial in terms of influencing and educating children about the environment than *what* is taught. The environmental problem needs to be offered as an issue open to debate. Children should be encouraged to develop their own interpretation of the problems and the potential solutions. To do otherwise is ultimately undemocratic and counter-productive. It is a delusion to believe that we can compartmentalise the different areas of our lives as if no interaction occurs between them. It would be rather hypocritical to advocate democratic ideals in certain parts of the curriculum (for example, 'education for citizenship') if undemocratic approaches to the solution of our local, national and global problems are practised elsewhere.

Our concern about the environment has never been greater, especially in respect of the pollution of the environment (Department of the Environment 1992). Concern is greatest among the younger sections of the population, but is matched by a sense of powerlessness and an inability to turn concern into action (Jensen 1993). Environmental education, however, often only serves to increase the sense of action paralysis among the population. One of the reasons for action paralysis resides in the status and mystique which surround science in an increasingly technologically dominated society (Breakwell 1989). Such status and mystification have been transferred to environmental education. With an emphasis on scientific, especially chemo-biological, investigations, children are taught that we have serious environmental problems. This is realised by giving children scientific information about environmental issues and providing them with experiential encounters with nature. It is frequently assumed that such information and experiential interventions will lead to significant changes in children's environmental values and actions. The function of these scientific investigations is to make concrete and imaginable the seriousness of the problems.

This approach, however, often only serves to reinforce feelings of powerlessness. Measuring, for example, coliform bacteria levels in streams as a consequence of agricultural pollution may serve to strengthen knowledge about the problem, but if it leads to a sense of helplessness and impotence the consequence in the long term can be a reduction in environmental concern. Furthermore, it can result in the individualisation of the causes of problems, and consequently the individualisation of the responsibility for them. The individualisation of cause and responsibility fails to set the problem within its wider social and economic context and disenfranchises anyone other than the farmer or agrichemical companies from responsibility or action. This simplistic understanding of environmental issues combined with a feeling of psychological estrangement might in the long term ultimately threaten democracy.

Feelings of powerlessness are further reinforced by the seeming remoteness of environmental problems. Empirical research we have undertaken in the UK, Ireland, Slovakia and Australia has clearly demonstrated that people see (global) environmental problems as more serious elsewhere and increasing in seriousness the farther they are away from them.

In support of these arguments we present the findings of a study assessing the effectiveness of an experiential environmental education programme.

A STUDY OF THE EFFECTIVENESS OF EXPERIENTIAL ENVIRONMENTAL EDUCATION

These problems are exemplified in a study which we undertook in 1993 in order to assess what children learn from a hands-on experiential encounter with the environment. The children who participated in the study were 63 female Year 10 students (mean age 14 years 4 months) from an Inner London school with a mixed catchment area. They were completing levels 4 to 8 of Attainment Target 2, 'Life and Living Processes', of the English National Curriculum at the time of the study (National Curriculum 1991). The children attended Sayers Croft Field Centre at Ewhurst, Surrey in England, between 7 and 11 December 1992. Sayers Croft is an Established Field Centre which uses both classroom and field-based environmental education methods to teach scientific topics covered in the National Curriculum.

Three groups of subjects were used in the study. The 'before' group (n = 32, mean age 14 years 3 months) included children who were tested before starting the field course. The 'after' group (n = 31, mean age 14 years 3 months) consisted of children tested after completing the field course. The 'follow up' group (n = 25, mean age 14 years 5 months) comprised children tested six weeks after the course, and included 15 subjects randomly selected from the 'before' group and 10 subjects randomly selected from the 'after' group. This was not intended to be a representative sample of all children and schools or indeed environmental education programmes. We believe, however, that the programme was typical of the content and teaching style of environmental education in England.

The principal aim of this study was to explore differences in young people's perceptions of the severity of environmental problems at both local and global levels before, after and six weeks following an environmental education intervention. It was hypothesised that children's perception of the severity of environmental problems at a local level would be higher in the after group compared with the before group, but lower in the follow-up group compared to the after group. Secondly, it was hypothesised that the children would assess environmental problems as more serious at the global level than the local level.

The same questionnaire was administered to all three groups of children. The questionnaire included several sections which focused on specific issues of environmental concern. In this chapter we focus on the assessment of the young people's perception of the severity of different environmental problems. The questions were partly derived from a questionnaire first developed by Woods (1991) which measured the perceived severity of different environmental problems at different areal levels. Four areal levels were used in the present study. These were 'you', 'your town', 'Britain' and 'world'. Six environmental problems which the children were covering in school at the time of the study were used in the research: water pollution, deforestation, the destruction of the ozone layer, overpopulation, air pollution and the disposal of nuclear and chemical waste.

The children were asked to rank the perceived severity of each environmental problem at each areal level on a scale from 1 (not serious at all) to 5 (very serious). The ratings of perceived severity for each environmental problem within a single areal level were added to give a perceived median severity score for that areal level. The median scores for severity across all the environmental problems at different areal levels for each sub-sample are shown in Table 12.1. This shows that there is a heightened level of concern for the different environmental problems with each increase in areal level.

Table 12.1 Perceptions of severity of the environmental problems at different areal levels: median scores

Group	'You' level	'Town' level	'Britain' level	'World' level
Before	20.50	20.00	24.50	30.00
After	23.00	21.00	27.00	30.00
Follow-up	18.00	18.00	23.00	30.00

Friedman two-way analysis of variance tests was employed to test for differences between the total areal level scores within each of the three conditions. The results of these analyses show that there were significant differences between the areal levels in all three conditions.[1]

Wilcoxon tests, which measure statistically significant differences in ordinal level data, revealed that there were no significant differences between the 'you' and 'town' levels in all three conditions. There were, however, significant differences between the 'town' and 'Britain' levels in all three test groups.[2] There were also significant differences between the 'Britain' and 'world' levels in all three test groups.[3] These differences are graphically displayed in Figure 12.1 which shows the median values for the total scores at each areal level for the three test groups.

The graph reveals two striking features. Generally, the post-intervention group registered a higher degree of perceived severity in the 'you', 'town' and 'Britain' areal levels than the pre-intervention group. This indicates that the children were more concerned about the specified environmental problems as a consequence of the environmental education intervention. However, six weeks after the intervention, their perception of the severity of these problems had declined to below the pre-intervention level. There were no statistically significant differences in the 'you' level between the before and after conditions or between the before and follow-up conditions. There was, however, a statistically significant difference in the 'you' level between the after and follow-up conditions.[4] However, it is clear from Figure 12.1 that the scores of the follow-up group revert to pre-intervention concern levels after arriving back at school (the follow-up median scores drop below the pre-intervention 'before' scores at the 'you', 'town' and 'Britain' levels). This suggests that environmental attitudes are fairly well entrenched. Furthermore, it suggests that what they learn on the course, both in the classroom and in the field, only serves to strengthen their views and perhaps heighten their sense of action paralysis.

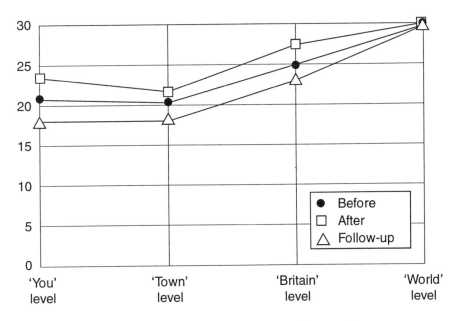

Figure 12.1   Perceived severity at different levels in the three conditions

DISCUSSION

Despite the emphasis on local experience and knowledge, children continue to have a greater appreciation of environmental problems at the global level and underrate problems at the local level. The maxim 'Think Globally – Act Locally' was coined many years ago in an attempt to overcome the phenomenon whereby the public become highly aware, largely through mass media images, of the destruction of exotic ecosystems and wildlife, but fail to appreciate that the same destructive processes operate on their doorstep.

The resolution of environmental issues at a local level through environmentally responsible and democratic means is considered a primary goal of environmental education (eg Stern 1992). From a pragmatic perspective, people have a better chance of influencing the outcome of local rather than global issues. Successful participation at a local level by well-organised groups has the potential to set a precedent and provide models of good practice for other groups to continue in this successful participatory process (Uzzell 1990). Our study, however, found that children were less concerned about problems at a local level than they were about problems at a global level.

This emphasis on global rather than local issues may be due to the fact that people's environmental belief systems are enmeshed within larger societal belief systems which may prevent the development of new beliefs (Uzzell 1992). Specifically, our societal belief systems tend to characterise environmental problems as being global in nature and not essentially located in the local, day-to-day world. This is reflected in the National Curriculum (1991) which heavily emphasises global and macro issues. Furthermore, research we have undertaken indicates that, not surprisingly, the mass media in general and television in particular are significantly more important than local information sources such as the school or the family. If children are to be encouraged to take action at the local level in respect of environmental issues and problems, then clearly they have got to see that there is a need for action at the local level (Uzzell et al 1994). One way of encouraging this is for the local mass media as well as schools to interpret global ecological processes in their local context.

The second important finding in our study was that there was a significant decrease in concern among the children once they returned home to the inner city. Evidence from this study suggests that environmental education as it is currently taught at one school and its associated field centre does not lead to a lasting change in children's environmental attitudes or values. It would have been very surprising if it had. If environmental perceptions and attitudes could be changed after a single intervention it would imply that such perceptions and attitudes are very superficial. It would also suggest that we could change people's behaviour more readily than we seem to be able to achieve at present.

This result might be explained in the following way. Science attempts to make the everyday things in life more unusual and to explain the commonplace in terms of abstract scientific concepts (Moscovici 1984). On the other hand, social representations are the result of our efforts to make usual that which is unusual. They aid our efforts to fit abstract ideas of science into a more concrete and familiar framework. According to Moscovici, social representations work in opposition to science.

The environment is the everyday world where these two universes of meaning interact. In this sense the representations of the world which we develop through the kind of environmental education currently taught in schools only serve to make unusual that which is usual. Environmental education ought to be the force which integrates them and makes them meaningful. Our research suggests that environmental education accentuates the two worlds and emphasises feelings of powerlessness. Returning to the school after an experiential encounter with the environment only further separates and compartmentalises these distinct experiences and different worlds – the world of the school and the scientific (abstract) understanding of the environment, and the world of our physical surroundings and the familiar (consensual and concrete) understanding of the environment. Consequently, children's concern decreases following an environmental education course, especially at the local level, because they cannot relate the scientific content of their lessons to the social world they inhabit. A greater emphasis on the application and implications of science is an obvious way of addressing this problem. Whether it is the most effective way educationally in the longer term is another matter. It might be more useful if children were encouraged and taught to interpret the everyday world in more abstract and theoretical ways. This places many more demands on both teachers and children, but there is little doubt that it would be a more powerful way of addressing this disjunction in the longer term.

Previous research by Fazio and Zanna (1981) showed that direct experience with an attitude object leads to stronger attitudes compared with indirect experience. Our study illustrates that a child's hands-on experience of the environment does not necessarily have this effect because the child does not acquire a hands-on experience of the environmental problem itself. The object of the educational exercise is, for example, not just to understand the chemo-biological causes of stream pollution but the clash of interests in society which regards the polluted stream as a problem. The problem is in society, not in the environment. A hands-on experience is invariably contextualised within a natural rather than social scientific framework. An 'experiential encounter' with the environment, such as measuring air pollution caused by cars or river pollution caused by agricultural run-off, invariably focuses on the symptoms of the problem, but rarely on the system and set of values which support one form of social or economic behaviour compared with another.

We do not experience the environment in subject or discipline-specific modes. An experiential encounter which focuses only on the chemo-biological world is as partial and deficient as one which solely focuses on the social and economic. An experiential approach within education to the environment has to address the social as well as the scientific. If, as was suggested at the outset, environmental education is ultimately to challenge and change the values, attitudes and behaviours of children as future stewards of the earth, then environmental knowledge has to be set within an appropriate social, political and economic context, the very one in which values, attitudes and behaviours are formulated.

NOTES

1. Before: $Xr2 = 63.27$, df = 3, p < .05; After: $Xr2 = 56.49$, df = 3, p < .05; Follow up: $Xr2 = 45.82$, p < .05
2. Before: 1 tailed W = -4.42, N = 30, p < .05; After: 1 tailed W = -3.91, N = 31, p < .05; Follow up: 1 tailed W = -3.68, N = 25, p < .05
3. Before: 1 tailed W = -4.78, N = 30, p < .05; After: 1 tailed W = -4.29, N = 31, p < .05; Follow up: 1 tailed W = -4.09, N = 25, p < .05
4. 2 tailed U = 247.5, N = 56, p < .05

REFERENCES

Breakwell, G. M. (1989) 'Young People's Responses to Scientific and Technological Change', Final report to the Economic and Social Research Council, Swindon.

Council for Environmental Education (1987) *Introducing Environmental Education, Book 2, Schools: Educating for Life*, CEE, Reading.

Council of the European Community (1988) *Environmental Education*, Resolution of the Council and the Ministers of Education, 88/C177/03, 24 May.

Department of Education and Science (1967) *Children and their Primary Schools (The Plowden Report)*, HMSO, London.

Department of Education and Science (1988a) *Education Reform Act*, HMSO, London.

Department of Education and Science (1988b) *Environmental Education 5–16, Curriculum Matters 13*, HMSO, London.

Department of Education and Science (1988c) *The Curriculum from 5–16, Curriculum Matters 2*, HMSO, London.

Department of the Environment (1990) *This Common Inheritance: Britain's Environmental Strategy*, HMSO, London.

Department of the Environment (1992) *The UK Environment*, HMSO, London.

ESRC (1992) *Environmental Education Research*, Economic and Social Research Council, Swindon.

Fazio, R. and Zanna, M. (1981) 'Direct Experience and Attitude–behaviour Consistency', *Advances in Experimental Social Psychology*, **14**, 161–202.

Hale, M. (1990) 'Recent Developments in Environmental Education in Britain', *Australian Journal of Environmental Education*, **6**, 29–43.

IUCN (1970) *International Working Meeting on Environmental Education in the School Curriculum, Final Report*, September, IUCN, USA.

Jensen, B. B. (1993) 'The Concepts of Action and Action Competence', paper prepared for the First International Workshop on Children as Catalysts of Global Environmental Change, University of Surrey, 3–4 March.

Moscovici, S. (1984) 'The phenomenon of social representations', in R. Farr and S. Moscovici (eds) *Social Representations*, Cambridge University Press, Cambridge.

National Curriculum (1991) *Statutory Orders: Science in the National Curriculum*, HMSO, London.

National Curriculum Council (1990a) *Curriculum Guidance 3: The Whole Curriculum*, NCC, York.

National Curriculum Council (1990b) *Curriculum Guidance 7: Environmental Education*, NCC, York.

Palmer, J. and Neal, P. (1994) *The Handbook of Environmental Education*, Routledge, London.

Schools Council (1974) *Project Environment*, Longman, London.

Scottish Education Department (1974) *Environmental Education: A Report by HM Inspectors of Schools (The Gilbert Report)*, HMSO, Edinburgh.

Stern, P. C. (1992) 'Psychological Dimensions of Global Environmental Change', *Annual Review of Psychology*, **43**, 269–302.

Sutherland, D. S. and Ham, S. H. (1992) 'Child-to-parent Transfer of Environmental Ideology in Costa Rican Families: an Ethnographic Case Study', *Journal of Environmental Education*, **23**, 3, 9–16.

UNCED (1992) *Agenda 21: Programme of Action for Sustainable Development*, The United Nations Conference on Environment and Development (3–14 June 1992, Rio de Janeiro, Brazil), United Nations Publications, New York.

UNEP/UNESCO (1975) *The Belgrade Charter: a Global Framework for Environmental Education*, UNESCO, Paris.

UNESCO (1977) *Final Report: First Intergovernmental Conference on Environmental Education*, Tbilisi, UNESCO, Paris.

UNESCO (1987) *Environmental Education in the Light of the Tbilisi Conference, Education on the Move*, UNESCO, Paris.

Uzzell, D. L. (1990) 'Triggering Behavioural Response', ESRC Workshop in co-operation with the Department of the Environment on Social Science and the Response to Climatic Change, Royal Commonwealth Society, London, 9 March.

Uzzell, D. L. (1992) 'Les approches socio-cognitives de l'evaluation summative d'exposition', *Publics et Musées*, **1**, 1, 107–23.

Uzzell, D. L., Davallon, J., Bruun Jensen, B., Gottesdiener, H., Fontes, J., Kofoed, J., Uhrenholdt, G. and Vognsen, C. (1994) *Children as Catalysts of Environmental Change*, Report to DGXII/D-5 Research on Economic and Social Aspects of the Environment (SEER), European Commission, Brussels.

Woods, V. (1991) 'Local, National and Global Environmental Problems: A Cross-Sectional Study of Irish People's Perceptions', unpublished MSc dissertation, University of Surrey.

World Commission on Environment and Development (1987) *Our Common Future*, Oxford University Press, New York.

13 Education for Environmental Responsibility: An Essential Objective

DOMINIC DIBBLE
University of Stirling

A curriculum which acknowledges the social responsibilities of education must present situations where problems are relevant to the problems of living together, and where observation and information are calculated to develop social insight and interest. (Dewey 1966)

INTRODUCTION

The above quote by the philosopher and educator John Dewey sets the tone for this chapter. I will shift the focus of Dewey's insight, however, by exchanging environmental for social in both cases, and by labelling the resultant attitude of environmental insight and interest which the curriculum is intended to produce as environmental responsibility. My aim is also more modest, in as much as I intend to describe not a whole curriculum, but merely the structure of a course which I believe should be a core requirement in any educational programme. In the course of this chapter I will be focusing mainly on the tertiary education sector. However, the suggestions I put forward could and should be adopted, with suitable modifications, within the secondary and even the primary sector. Indeed, the engendering of an attitude of environmental responsibility should ideally start as early as possible.

DO WE NEED IT?

Two questions immediately arise when considering the development of an independent course on environmental values. First, is a separate course specifically designed to foster an attitude of environmental responsibility necessary? Surely we can trust the individual (as we tend to do in the case of social responsibility) to pick up an appropriate range of attitudes from the courses which he or she already takes? This raises the second question of whether an attitude of environmental responsibility is in fact normally implicit in the structure of academic courses, and whether it is in principle possible to deliver a coherent and consistent account of such an attitude through the mediating filter of different academic disciplines.

Values and the Environment: A Social Science Perspective, Edited by Yvonne Guerrier, Nicholas Alexander, Jonathan Chase and Martin O'Brien. © 1995 John Wiley & Sons Ltd.

In response to the first query, I argue that environmental responsibility is not clearly implicit in many, if not most academic courses. This is perhaps not very surprising, as the environment has only recently become a topic of wide concern and interest, whereas our modes of social behaviour have been of central importance to us ever since we began to organise in social groupings. So this provides one good reason for thinking that a course designed specifically to address the issue of environmental responsibility is an important and necessary addition to the curriculum, as we do need to examine this issue urgently, and it is a little ambitious to expect the structure of courses in every discipline to change overnight to accommodate this fact. Furthermore, there are some courses which seem poorly suited to the incorporation of an environmental subtext – for example, pure mathematics, theoretical physics, and perhaps to a lesser extent, the study of languages. It is therefore important that those who follow such lines of enquiry have access to some means of reflecting on this issue, as it is one to which everyone should give serious consideration. Thus an individual course which is a core requirement for all in education would seem to be indicated as an interim measure, until its message has been fully incorporated into the various different ways in which we analyse the world.

In response to the second query, it is important to acknowledge the impact on environmental understanding of the disciplinary organisation of knowledge. One characterisation of this organisation which is fairly clear is provided by Daly and Cobb (1990). They indicate that this organisation requires that each discipline have a subject matter clearly distinguished from the others, that the definition of a discipline also requires methodological self-consciousness, and that the method must be one which not only illumines the separated subject matter, but further selects the features of the subject matter that will be noticed and treated. Addressing the discipline of economics, Daly and Cobb note that these characteristics limit the number of people who may call themselves economists and who receive a salary as an economist (Daly and Cobb 1990, p33). For economist, we can substitute the title of any other type of academic, eg physicist, historian, etc, etc.

One thing that comes through quite plainly is the idea of *separation*: each discipline has its own area of expertise and techniques for examining that area marked off from all the others. But it is precisely such clear divisions between areas of knowledge which create significant impediments to the free flow of ideas about the environment – in other words, the relative impermeability of disciplinary boundaries erects intractable barriers to the development of knowledge about the environment. The reasons for this are complex and are beyond the scope of this chapter; for an extended discussion, see Daly and Cobb (1990), particularly Chapter 6. Nevertheless, it seems evident that a trend which leads towards a separation of disciplines from one another will tend to impair their ability to work together on problems which cross disciplinary boundaries; and environmental problems are paradigmatic examples of these. Furthermore, the interdependence and interconnectedness of environmental systems pose serious questions about the whole idea of making sharp divisions between bodies of knowledge to produce relatively autonomous and independent disciplines.

Returning to Daly and Cobb's point about the right to be titled an economist (chemist, musicologist etc), it can be argued that this right is reinforced by the way in which universities and other academic institutions are normally organised into departments centred on particular disciplines. Given the fact that in the present climate the various departments must compete with one another for resources, there is an inevitable temptation to champion one's own discipline as of the greatest relevance to a particular problem or set of problems, for reasons which have little or nothing to do with the desire genuinely to increase understanding. This point is particularly relevant with regard to the environment, as there is currently a considerable amount of research money available for projects which address environmental problems.

On the basis of these preliminary observations it seems obvious that an attitude of environmental responsibility is not something which we should expect people to learn almost subliminally from the courses they currently take. Therefore, I believe that there is a *prima facie* case for a stand-alone course which is a core requirement for all. The following remarks form an outline of one such course.

PRINCIPLES OF ENVIRONMENTAL LEARNING

Underlying my proposals for an independent course in environmental responsibility are three important educational principles. These are that such a course:

1. needs to deal with both facts *and* values;
2. should be largely user driven;
3. should be a compulsory core requirement for all students, while at the same time being non-assessed.

I now deal with each principle in turn.

FACTS AND VALUES

The first principle stems from the fact that such a course is concerned not simply with the environment as an object of supposedly dispassionate study, but with how we have *chosen* to study it, and therefore with how we currently relate to it and with how that relationship may need to change. It seems clear that the ways in which we relate to something are determined both by facts about that particular thing and by how these facts relate to our values. It is therefore essential that both aspects of the relationship are addressed if we are to obtain a balanced picture. Given this requirement, there arises the practical question of course leadership: who would be best suited to lead the course? Since a broad range of facts is likely to be examined, there is a need to ensure the involvement of someone with training in techniques of environmental research: an environmental scientist or geographer, for example. On the other hand, the question of 'values' arises most consistently in the discipline of philosophy and related fields of law, anthropology, psychology and sociology. Hence my suggestion is that the course should be jointly led by a suitable combination of two (or more) tutors who can deal with both aspects. Although implementing such joint leadership might lead to scheduling problems on the part of the academics involved, there are available a number of technological solutions to the difficulties of ensuring widespread access to the course materials. In particular, the development of information technology and computer-assisted learning packages means that staff and student engagement with the course materials can be organised efficiently and with less disruption to timetabling schedules than might otherwise be the case. I return to the issue of software tools below when I describe the conduct of the course in more detail.

A STUDENT-LED COURSE

The second principle is closely connected to the first, in as much as part of the aim of the course is to make explicit how all those involved relate to the environment. This cannot be done by the students only being presented with a series of facts and arguments by the course leaders: all this will do is (implicitly) indicate how the course leaders relate to the environment. If the course is to produce any significant change in the attitudes of those present, then it must flow from their own present knowledge and attitudes. This implies that the actual progress of the course will be unpredictable in advance and will differ from group to group. Indeed, a comparison of the different directions which different groups take would serve to underline the complexity of the environment and the diversity of opinions which people hold concerning it. This would provide further evidence to support the idea that no one discipline has the right answers in this field. In these senses, the course encourages/represents the development of a community of enquiry which attempts to follow the enquiry where it leads rather than being penned in by the boundary lines of existing disciplines (Lipman 1991, p15). So as long as all those involved are willing to contribute to the discussion, they will almost certainly find that the issue of the divisions between disciplines will arise naturally (again, a point to which I return briefly below).

A CORE, NON-ASSESSED COURSE

In the section on why education for environmental responsibility is necessary, I gave a general theoretical justification for the development of a stand-alone core course dealing with environmental responsibility. There are also two practical justifications. The first is the rather obvious one that every student, like every other member of the population, has to make decisions which affect the environment; therefore every student should be given the opportunity of making those decisions on a more informed basis. The second returns to the issue of disciplinary barriers, for if there is a mix of students from different disciplines within each group on the course, then it seems even more likely that they will soon discover both the positive aspects and the shortcomings of their own and other disciplines in dealing with the environment. With regard to the strategy of non-assessment, there are three good reasons for this step. First, and perhaps most importantly, the course is liable to focus more on the discussion of attitudes than of facts, and if successful, will affect the attitudes of those involved. But who is to decide what is the correct change in attitude to be achieved by the course? Constructing a standard attitude would amount to indoctrination, and hence would negate one of the central aims of the course – adopting a more reflective stance with respect to different *types* of doctrine. Secondly, given the somewhat unfamiliar nature and conduct of the course, students may feel inhibited and the knowledge that anything they do contribute will not be taken down in evidence, and that they do not have to strain to listen out for the questions which will come up in the exam, will help to counter-balance inhibition or nervousness. This argument takes into consideration evidence that learning is facilitated in a low-threat environment (cf eg Gibbs 1981, pp89–90). Finally, since different groups would proceed in different directions, it would be impossible to set a standardised exam.

CONDUCT OF THE COURSE

In order to clarify just how such a course might work, I will sketch out some initial ideas on its practical implementation. Since discussion would be the major activity, this means that the numbers involved in each group would have to be fairly small – say 20 students. This indicates that mounting such a core course may produce organisational problems. An educational institution with 5000 or 10 000 students faces immediate practical dilemmas about timetabling, the provision of appropriate meeting space and a sufficient range of learning resources. However, it is crucial that a commitment to environmental responsibility is developed in contemporary educational institutions; if the environment is to be addressed seriously then resources need to be channelled so that practical problems can be overcome.

One option for helping ease such difficulties would involve the intelligent use of computers. Software packages which provide conferencing facilities offer the possibility of conducting tutorials asynchronously, ie the tutors could place the relevant information/questions on institutional computer networks, and over the course of the week students could come in and make their contributions to the discussion and analysis at differently scheduled times. The main advantage that such packages have in this respect is that they would in theory allow the students to survey easily what contributions others had made, and to link their contributions to them, either in support or magnification or disagreement – thus there would arise the possibility of a genuine debate between students. The asynchro-

nous nature of the debate might be helpful to tutors as well in helping to sur-
mount scheduling problems, since they would not have to be in the same place at
the same time, but could input their contributions as and when necessary.

Of course, any use of information technology in this manner needs to be
conducted self-consciously. Information technology itself and its use by large
numbers of learners generates environmental impacts. The impacts of the learning process and the technologies through which learning is organised also need
to be a part of the critical reflection that a course in environmental responsibility addresses. Ideally, given sufficient staff, there would be no need for its use –
there is, however, no guarantee that large numbers of staff would be available.
Therefore, just how large a part the use of such software would play in the
course would be a matter for experiment and negotiation. I am not suggesting
that it should fully replace normal tutorials and lectures, but that its intelligent
use could substantially facilitate the conduct of the course. Thus in what follows, interactions between tutors and students and among students may be
taken as happening *either* in the conventional manner *or* through the medium of
a software package on a computer. If very little or none of the significant work
is done on computers, then one way to provide some concrete record of the
group's endeavour would be to have one person acting as a secretary (perhaps a
different person each week, perhaps one of the leaders). The secretary would
record a brief summary of what was discussed at each meeting, and at the
beginning of each meeting the summary of the previous meeting could be
issued. Then, at the end of the course, the course leaders could prepare a group
document based on these summaries.

STRUCTURE OF LEARNING

The course can be organised in terms of a small number of simple stages:

1. *Preliminaries*:   the tutors lay down the ground rules concerning their role
   and the role of the students in proceedings.
2. *Generating initial questions*:   tutors and students collaborate to discover
   questions which will act as an appropriate starting point.
3. *Clarifying initial questions*:   the initial questions are refined through dialogue among the participants.
4. *Answering initial questions*:   the questions are debated until the group is
   satisfied that the question has been examined as exhaustively as is desirable;
   this examination may or may not be facilitated by additional factual information from the tutors, depending on its subject matter. In the process of
   debating the initial questions, new questions will almost certainly have
   arisen: hence the next stage is to return again to stage 3 and continue.

The process has no formal conclusion; its objective is precisely to encourage
a persistent questioning attitude and commitment to evaluating social action in

environmental terms. The 'end product' of the course is the ability to generate questions about the environmental consequences of different types of action. Hence, the four stages are intended to facilitate the development of a questioning *process* and an evaluative *attitude* to the world of experience. I will now comment briefly on each of these stages.

PRELIMINARIES

Course leaders must make the nature of the course (particularly its user-driven element and the lack of assessment) and their own role within proceedings clear so that the students realise their responsibility for playing an active part in the conduct of the course. Actually, the role of the course leaders is somewhat difficult to define – they must strive to avoid imposing their own attitudes or force-feeding the students with unwanted ideas, while at the same time seeking to keep the discussion from drying up or running down blind alleys. Thus their role is akin to that of a team coach, with a sensitive awareness of individuals' abilities and needs within the group and the motivational skills required to encourage the demonstration of these abilities. They will also act as sources of information on their areas of expertise, and as guides to sources of information outside their discipline, but they will have to exercise discrimination as to when they provide information. There may be some occasions when information should be withheld even when requested, and others when it must be provided in order to prevent the discussion from straying into major error.

GENERATING INITIAL QUESTIONS

Having taken care of preliminaries, those involved must get down to the business of generating the questions which will guide the progress of their enquiry. This could be done in a number of ways:

1. Course leaders could simply ask the students to produce a list of factual questions and a list of value questions which they think that their discipline and other disciplines raise.
2. Course leaders could be slightly more directive, and provide the students with a number of key concepts, such as environment, value, environmental impact, responsibility and academic discipline, and ask the students to generate questions based on these terms.
3. Course leaders might decide to begin with some leading questions, such as 'What is the environment?', 'Should we value the environment? If so, how do we go about it?', and invite the students to use these as starting points for generating their own questions.
4. A short text on an environmental topic could be distributed for examination and comment.

Since the first suggestion gives the students the most responsibility and lee-way, it would seem on the face of it to be preferable, but there will be cases where a particular group needs more guidance in the initial stages. The process of learn-ing to formulate your own questions to guide a particular line of enquiry is one which may be relatively unfamiliar to the participants; it is an area of education to which more attention needs to be given and it will almost certainly require con-siderable skill from the tutors to evoke an initial set of questions which will be sufficiently rich in content to generate a useful and stimulating debate.

CLARIFYING THE QUESTIONS

Once the initial questions are in place and the group has decided which ques-tions need to be addressed first, then discussion and analysis can begin. Probably the first issue which will arise in the discussion is the need to clarify and make more precise some of the questions. For example, the question, 'Why are we doing what we're doing to the world?' could receive the simple answer, 'Because we're greedy and short-sighted'. However, this does not really do jus-tice to the complexity of the problems, so it would be better to analyse the origi-nal question, asking, for example, 'Who is meant by "we"?', 'Just what *is* it that is being done?', 'What is intended by "the world" here?'. If the question has already received the above answer, then the answer can receive similar treat-ment, eg 'Why are we greedy?', 'Is greed good?' etc. The course leaders must be careful not to produce the questions themselves, but must try to get the students to recognise the complexity and uncertainty of the original question and answer.

If the course is to make any significant progress in examining people's atti-tudes, it will be important to analyse what people think about specific concrete situations; and to do this, it will be necessary to learn how to refine more gen-eral questions. Good examples of how to proceed in leading a group of students to refine and extend their questions are provided in Postman and Weingartner (1971, pp75–9, 110–12). During this process it may emerge that a deeper exploration of a given concept (eg ecosystem, contingent valuation, ecofemi-nism etc) with which the students are unfamiliar would be helpful. The course leaders may decide that a lecture session on this concept is appropriate and, with the students' permission, provide it. Of course, if one of the students is familiar with the topic and feels comfortable about sharing his or her know-ledge, then it would be preferable for that student to take responsibility for pro-viding the relevant information and analysis.

ANSWERING THE QUESTIONS

Once the group is content that the questions are precise enough, then the enquiry will flow from the group's attempts to find satisfactory answers to these questions. This will require an articulation of how the participants inter-pret the questions in the light of their own individual attitudes; the validity of these attitudes, and the preconceptions on which they are based, will then be tested in the crucible of debate. In the light of this testing procedure, possible routes to an answer which will satisfy the group should become clearer. But is this enquiry to remain uninformed by past thinking? Should there not be some written material which can be used to guide the enquiry? This is an interesting issue – great care must be exercised with regard to the issuing of written material in order to avoid prejudicing the direction of the enquiry. At the same time, it is certainly useful to discover what others have thought. One way to deal with this is to explore as far as possible what the group thinks on a particular question, before recommending relevant written material. The group can then compare what others have thought with their own reflections, and can discuss the strengths and weaknesses of both. Once the group is satisfied that a particular question has been sufficiently discussed, then they can move on; and of course, as the enquiry proceeds, new questions will be generated.

FINISHING THE ENQUIRY

As the enquiry has no formal goal, it would end when the course ends – a course which is one term or one semester long would probably be sufficient to sow the seeds of future reflection in people's minds. The outcomes of the students' discussions and investigations can be made available in a number of ways:

1. Where specific software packages have been used, a hard copy of that part of proceedings can be made and distributed among the group.
2. Where a secretarial role has been established, the notes generated by the group can be collated and distributed.
3. Finally, students can be encouraged to keep reflective journals in which they maintain a record of their impressions of the course contents and a note of any changes in personal philosophy and behaviour which have resulted.

In both the first and second cases, the hard copy of the students' work could be made available to students in subsequent years. This would act as a reference source. Thus it would give future students an insight into how the process of generating and answering questions works; and by indicating what topics were of concern to their predecessors, it would give them the option of exploring these topics in more depth, or of exploring fresh areas of concern. In the third case, students would acquire experience in maintaining reflective journals and would carry into their future careers their own personal ways of making sense of environmental questions. These reporting methods can be used either separately or in combination.

CONCLUSION

The implementation of such a course constitutes quite a challenge. Mounting it would at first be something of an experiment, with appropriate modifications in design being made in response to its first participants; and it might be difficult to persuade others in the university of its importance, leading to problems in acquiring sufficient resources (the departmental/disciplinary structure of the university might prove a hindrance here – eg which department(s) would fund it?). The major objective of this course might be characterised as to produce the conditions in which attitudes concerning the environment are able to change. But is there any need to change attitudes? Do we not know enough already about the environment? As I remarked earlier in this chapter, a great deal of our knowledge tends to be filtered through the distorting lenses of the different disciplines: this course should at least enable those who take it to recognise that those lenses *do* come between us and the world, and that they *are* distorting. It is this recognition which will give those involved the courage and insight to change both their intellectual and moral attitudes. I believe that this ability to change is itself of crucial importance. To quote Carl Rogers:

> The only man [sic] who is educated is . . . the man who has learned how to *adapt and change* . . . Changingness, a reliance on *process* rather than upon static knowledge, is the only thing that makes any sense as a goal for education in the modern world. (Rogers 1983, p120, first emphasis added)

If a course like the one proposed here can indeed give those involved the ability to change, then it will stand them in good stead in every area of life, including their role as citizens, for the ability to change their attitude to a given situation is required before they can change their *response* – in other words, become more *response-able*. This is particularly important given the rapid rate of change taking place in the physical and socioeconomic environment of the modern world. How can people take environmentally responsible decisions if their responses to the environment are dogmatically fixed in patterns created in the past? Citizens whose minds and hearts are open to change are much less likely to fall prey to political dogma and social prejudice, and much more likely to be able to take truly responsible decisions – *environmentally* responsible decisions – concerning both their social and physical environment.

ACKNOWLEDGEMENTS

I would like to thank Fiona Tilley for her comments on assessment, and Martin O'Brien for his suggestions.

REFERENCES

Daly, H. E. and Cobb, J. B. (1990) *For the Common Good*, Green Print (imprint of The Merlin Press), London.
Dewey, J. (1966) *Democracy and Education*, Free Press, New York.
Gibbs, G. (1981) *Teaching Students To Learn*, Open University Press, Milton Keynes.

Lipman, M. (1991) *Thinking in Education*, Cambridge University Press, Cambridge.
Postman, N. and Weingartner, C. (1971) *Teaching as a Subversive Activity*, Penguin Education, Harmondsworth.
Rogers, C. (1983) *Freedom to Learn for the 80's*, Merrill (imprint of Macmillan Publishing Company), New York.

14 World Views and Environmental Action: A Practical Exercise

MARTIN J. HAIGH
Oxford Brookes University

INTRODUCTION

Environmental action is affected by the perceptions, value systems and structures of economic and political power and privilege within a community. Environmental conflicts and controversies often arise from the clash of communities, and power blocs within communities, which hold fundamentally different views about the nature, significance and value of environmental attributes. These issues become more critical when they involve interfaces between different cultures. Frequently, these differences in viewpoint are much more fundamental than anything implied by the term 'value system'. The differences at the core of much environmental controversy often involve matters which the participants regard as absolute, self-evident and unquestionable truths.

The ways in which different groups ascribe worth to different aspects of the environment is prejudiced by preconceived notions. Further, different groups and individuals may apply completely different rules to the same situations, depending on the context and their perceived role in that context. Whether the environment is seen as a resource to be developed, an heirloom from previous generations, a precious natural habitat, a workplace for people, or a gift from God which must not be despoiled, such preconceptions are the foundation stones on which both public and private values are constructed.

Arguably, one of the most important of the 'personal transferable skills' which may be gained through education is an ability to understand and appreciate the viewpoints of others. Students who, on graduation, may become actively involved in environmental action, or controversy relating to such action, need some appreciation of the nature, origin and provenance of different ways of conceiving that action. The fact that our higher education system still produces graduates who neither understand that others may validly conceive the world differently, nor yet guess how those others may perceive them, ranks among its greater defects. Releasing such individuals into careers where they make decisions affecting the environment is a certain route towards perpetuating the present dismal spiral of environmental conflict and disruption.

Values and the Environment: A Social Science Perspective, Edited by Yvonne Guerrier, Nicholas Alexander, Jonathan Chase and Martin O'Brien. © 1995 John Wiley & Sons Ltd.

BACKGROUND: RAISON D'ÊTRE

The starting point for this exercise is a belief that university education is too focused. University teaching tends to operate across a relatively small range of rationalist and secular world views. Much is dominated by the precepts of science, especially the reductionist scientific method, technology, where mathematical argument is leavened by the cookbook recipes of experience, the laws of economics, the case book of business studies, or by the literary and philosophical underpinnings of white, Western, ultimately European society with its cultural aspirations and political thinking.

Even university social science has lost contact with the outside world. It remains a place where subtle differences in the ideas of 'great Western thinkers' matter more than the huge differences in the beliefs of sections of current society. Meanwhile, the ideas from those majority of nations and social groups who lie beyond the pale of European and American respectability are shunted away for specialist study in small, underfunded, 'minority' and area-study programmes.

Outside, the world has changed. Despite the economic and political dominance of the West, and the remaining influence left to the residuum of its socialist antithesis, non-Western societies are increasingly taking over economic, technological, political and, ultimately, cultural leadership in the world. Even in the West, the pale is breached. Western societies have become ethnically and culturally more diverse and the new communities are demanding that their views are noted. As the old faiths in the monoliths of church, science, Marx, nation state, and the commercial/industrial progress of social Darwinism fade away, new ideas and new beliefs move in to fill the vacuum. Since the outside world is no longer such an unfamiliar and inaccessible place, many come from outside. So, while Lord Krishna may not receive much attention at university, he is quite at home driving his chariot across posters in the local primary school, in the London Underground, and in the Moscow metro.

The conceptual turmoil is the greater because Western thinking itself may be in the throes of a major paradigm shift (cf Kuhn 1970). From small mutterings of concern in the 1920s, the noise of protest against the habit of thinking of the world in inanimate, mechanical terms is becoming very loud. Some speculate that this groundswell is evidence that the West is shifting towards a phase where thinking is dominated by concepts of ecological self-sustainability rather than economic expansion (Metzner 1993; Capra 1982).

The exercise described in this chapter aims, in some way, to loosen some of the blinkers imposed on students by university thinking and, at the same time, provide a glimpse of the possibly imminent transformation of social and environmental values. The educational problem is how to present this pluralism, this dynamic and this general turmoil in a way which may be helpful to a student who has no special interest in or knowledge of such matters, and indeed who may be intensely preoccupied with the imminent launch of a personal career in 'tourism and hotel management' or similar.

THE LANDSCAPE ASSAY

The 'Landscape Assay' is a field study programme designed for upper-level undergraduates and estimated to require around 30 hours of average student effort. The programme has three aims. First, it attempts to remind students of some of the dimensions and implications of social, cultural and political pluralism. Secondly, it tries to immerse students in what it actually means to hold a particular world view which is alien to them. Finally, it tries to build up the students' abilities to listen to others and detect the cues which indicate the influence and characteristics of an enfolded value system.

The programme tries to achieve its aims by the active involvement of participants in a series of reading, discussion and simulation exercises. In brief, students form into teams, read up and assimilate a particular viewpoint on the environment, then apply this viewpoint to the interpretation of a part of the natural environment, in this case a reach of the River Thames in Oxford. Each team is asked to report back on what is good, what is bad and what is otherwise in the landscape. They report back in three ways. First, they prepare written reports which describe their interpretation of the philosophical/ideological system which they have been assigned and a description of how they used this to interpret the environment. Secondly, they make a personal, usually spoken presentation to the whole class which concerns their findings, but which does not describe the system which guided their approach. Finally, they try to determine, and discriminate between, their system and the core beliefs and derived values which guided the reports of all the other teams presenting to the class.

SYLLABUS

The syllabus spans three main systems of human thought. These include systems based on internal, spiritual or subjective revelation, systems based on the analysis of the external, material/mechanical and objective, and systems based on holism and the recognition of emergent properties in evolving living entities (Haigh 1993, but cf Rorty 1980, pp15–70; Shiva 1993). The three systems are introduced through discussion of the nature of metaphysics and the form of a metaphysical or 'world hypothesis' (Pepper 1942; Van Inwagen 1993). The author shares with Pepper (1942, p1) and Rorty (1980, p12) the view that 'it is pictures rather than propositions, metaphors rather than statements, which determine most of our philosophical convictions'.

In this case the three images are the mind (or perfect body), the organic living body, and the mechanical, inanimate, material body of the external world (cf Haigh 1993). These are presented via a specially created workbook in the form of a Hegelian dialectical spiral. The narrative begins with the spiritual systems, continues through discussion of their materialist antitheses, and ends with resolution in the syntheses of organicist holism (Haigh 1994). Intended as both remedial and revision material, the workbook tries to introduce core ideas and key readings from geography and environmentalism at large. Topics covered include some fundamental ideas from Hindu and Islamic (Sufi) thinking, followed by notes on Christian environmental ethics, animism, pan(en)theism, and the Lynn White contentions about the origins of our current environmental crisis (Pojman 1994). Sections on materialism begin with the Hindu Lokyata doctrine, then mention Descartes, mechanicism, Darwinism, Hegelian and Marxian dialectics, monetarism, contingent valuation and the scientific method. Finally, vitalism is introduced followed by formist idealism, modern organicism, ecologism, and the work concludes with the reification of Gaia (Haigh 1993).

The workbook is interspersed with exercises which ask the reader to use concepts as they are introduced. However, the last and only assessed question closes in to the central purpose by asking the student to explain the world view and environmental value system which is being illustrated by the workbook's author. To provoke student speculation, the Hegelian metanarrative is leavened with sections emphasising the importance of 'personal subjective revelation', 'evolutionary change', and both begins and ends with the image of Shiva (Natraj) dancing in the soul of the universe. A further clue is provided by repeated emphasis of the excellent critique of (and dire warning about) such thinking provided by Elliot Miller (1989). Nevertheless, student respondents often argue that the author may be a Marxist, a closet Hindu, or a woolly liberal environmentalist. At this point, the answer is less important than the fact that the student is trying to find that answer and attempt a critical deconstruction of another's motivation.

Students are then introduced to some generalised examples of topical world views through a second workbook on the evolution of geographical thought, a seminar, and through work on some supporting readings (Gold 1994; Pojman 1994; Tucker and Grim 1993). They are then required to move on to employ some of the world views suggested.

In one typical exercise, they are asked to produce contrasting interpretations of specific environmental events. One of these is based on a written report of a disaster caused by the failure of the Arkansas River railway embankment in 1927 (Box 14.1, cf Tagore 1965). This exercise examines world views covered in the section of the workbook devoted to spiritual systems. A sample of student responses is contained in Box 14.2.

A more elaborate second exercise follows. This uses extracts from the movie of Steinbeck's *Grapes of Wrath*, which concern the causes of the Dust Bowl in the Great Depression of 1930s USA (Gold et al 1995). The views examined during this exercise include environmental determinism, ideas from the geography-specific schools of environmental possibilism/perception, fundamentalist Christianity, Marxism, and Gaianism, all presented to the student via short written papers published by committed protagonists.

Box 14.1   Teaching students to appreciate different world views: extract from the 'world hypotheses' workbook

WH DISCUSSION TASK 1: There follows a brief description of an event. All you have to do is interpret this event. Several interpretations are possible based on the ideas described in the notes that follow the subhead: 'The Self in Its World'. Your task is to provide four, mutually exclusive, understandings of this event, its significance, and its implications.

This is what happened:

> This year, 1927, the warm spring rains began early. They started in mid-March while the earth was still frozen. The rain slid off the hard ground and found its way into rivers and creeks. For four weeks there was scarcely a day when the rain did not fall . . . On the Arkansas River, which drops with sudden great jumps and then slopes gradually, the tail of the flood caught up with the head. The flood level was the highest in 99 years. Its force smashed a flood protection levee and tore away 400 yards of earthen wall. A passenger train racing through the night hit the washout and pitched into a ditch. (Kerr 1960, p93)

Why did this happen? What should we humans do next?

Here is a clue to start you off. The text comes from the writings of a 'creationist'. It begins thus: 'a literal translation of the first few verses of Genesis has been provided to me by an outstanding scholar. It serves our modern scientists well in establishing their thesis' (Kerr 1960, p23).

Use 100 words (maximum) for each 'understanding'. Be prepared to discuss each of your viewpoints in class.

Next the students form into small teams, to research and apply a still more specific philosophy. This is indicated by a briefing sheet which does little more than offer a small number of illustrative quotations and a few references. The list of briefings, constrained mainly by the knowledge of the teachers involved, includes examples from all three of the grand metaphilosophical systems. Spiritual systems included are: mysticism (Buddhist), mysticism (Sufi – Islam), sarvodaya (Gandhi), aquarian (New Age panentheism), Christian (creationist, literal fundamentalist), Hindu (Vaishnavite), and Western romantic idealism. Briefs have also been prepared for the following organicist/vitalist systems: formism (classical vitalist idealism, New Age Sheldrake-Lamarckianism), Gaianism, and general systems theory.

Finally, briefings are prepared for some materialist/mechanist systems: Marxism (Communist Party of the former Soviet Union – official line), scientific ecology, landscape planning, river engineering, contingent valuation, environmental management auditing, and for some more overtly hybrid systems objective aesthetic evaluation and romantic ecologism. The guidelines for the exercise offered by the student handbook are shown in Box 14.3.

Box 14.2   Typical student responses to world views exercise

1. God has punished the impudence of those who sought to capture Her River and sought to constrain Her with puny banks of earth.
2. The River God and the Rain God danced together for joy at the birth of Spring. Unfortunately, a train got in the way.
3. It was God's will that the river should break its banks and preordained that the train's passengers should die in this manner at this time. The message of this verse in the Holy Book of experience will become clear upon reflection.
4. My mind has self-created some serious imperfections in my world. I must concentrate hard to eliminate these failings.
5. This event has been created for my personal instruction. I must find a way to respond to this message by altering my way of life and/or thinking.
6. Clearly, the passengers on the train had sinned and now they have been eliminated as an example to us all.
7. Those who were guardians of the river must strive harder in their duties which are to provide strong, flood proof banks for the river. No doubt God will punish them for their failings. I must strive to do my duties better and so avoid such a reprimand.
8. The battle to subdue the Earth continues. The evil forces have caused this setback. We will rebuild and struggle onwards.

PRACTICAL

This field study involves students, organised as small project teams, in applying a prescribed world view to the appraisal of a particular landscape. The student teams are asked to assay this landscape for what, according to the lights of their prescribed world view, was good, what was bad, and what was otherwise. As any gold digger knows, an assay is a series of tests which are applied to rock to determine what it contains which is of practical value. However, in this case the physical attributes of the target landscape provide no more than a source of illustration. The subject of the assay is not the 'landscape' but the world view or philosophical system which each student project team must attempt to understand and interpret. (Indeed, realising this, in 1994 several of the better student project teams found it inappropriate to visit the target field study area at all, while another went to a completely different location).

This revelation may be repeated incessantly to student participants in the programme. However, it is unlikely that it will register until the students are well into their work. This has important implications for the teacher. It is not enough to introduce the field exercise and let the students get on with it.

The exercise must be introduced in two stages. First, the aims and methods must be set out; then, when the student teams have done a little reading and are

Box 14.3 Sample landscape assay exercise team briefings

M2662.3: ASSAY OF LANDSCAPES ALONG THE RIVER THAMES

This exercise involves you in the practical application of a philosophical system. The exercise concerns landscape interpretation and the rules/screens/filters which societies/individuals employ to separate out what is important/valuable and what is unremarkable/value-less in the world which surrounds them.

This exercise requires you to build a methodology, a set of operating instructions, guidelines, and/or rules, which you will apply to a local landscape interpretation case study. You will classify the landscape attributes along the River Thames into three types: 'Good', 'Bad', 'Indifferent/Other'.

Your methodology will be based on the ideas contained by a written text or texts. This text or these texts will exemplify a way of seeing the landscape by providing a set of emphases, priorities and values. The reading(s), however, will not tell you how to do your appreciation. This is for you to discover through library research and your own collective imagination. Your task is to design a technique for landscape appraisal which is TRUE TO THE SPIRIT OF YOUR GIVEN TEXTS. You aim to create a methodology which is identical to that which the author(s) of your text(s) would have devised were they set to do this exercise.

---

about to embark on their field study, a second class workshop must be organised. The starting point for this is the question 'What is this exercise about?'

Traditionally, the first answers will be things such as 'landscape', 'landscape evaluation' or 'the River Thames'. However, once these naive and incorrect notions are discarded, attention can be focused on the student's immediate problem of turning an ill-defined set of ideas into a way of seeing the external world. Since these problems are often fairly specific to the particular brief, the detail of these problems is best reserved for tutorial meetings which address the plans of individual teams. The workshop, however, may be used to run through a worked example or case study, perhaps involving a world view not included in the briefings list.

Experience suggests that two tutorial meetings are usually needed to support each of the better project teams. The first tutorial meeting is typically devoted to discussing and refining the student team's methodology for landscape appreciation. This resolves into the issue of prioritising, *a priori*, what followers of such thinking might consider important or valuable and what unimportant or bad in the world. The second tutorial meeting more usually focuses on the student team's findings, and more particularly on how they may be presented to the class most appropriately. Teams considering more adventurous modes of presentation especially require reassurance.

RESULTS – THE FINAL REPORT SESSION

The final session begins with each team handing in its written report on the philosophy they have studied and the methodology they employed to create their assay of the River Thames landscapes. The class is divided into sets, each containing 5–6 teams offering a representative selection of world views from the spectrum of enfolded root metaphors.

The two-hour session begins with each team explaining to the whole class how they detected and differentiated between landscapes which are 'good', 'bad' or 'otherwise'. Teams are expressly forbidden, on pain of enormous grade penalties, from mentioning the subjects included in their written submission, especially the ideology which guided their classification. The reason for this is that, subsequently, the class is given the task of trying to interpret the belief systems which created each, deliberately very different and often mutually exclusive, assay.

Techniques employed for presentation have been reassuringly diverse. Occasional lecture/slide presentations have been mixed with a great deal of role playing, much re-enactment, much audience participation, much music, some dance, some mathematics and a little chanting. Furnishings have been refashioned variously as personifications of television studios, punts, laboratories, temples, party meetings and chaos. One presentation has been entirely unspoken – the team communicated with painting and music. One important function of the session is to encourage students to explore the validity and suitability of different communication media.

The session proceeds as follows. Prior to the class presentations, each student is given a work sheet and asked to record first 'According to group A, B, C, D, etc, what properties are 'good' in the environment?', and then, 'According to group A, B, C, D, etc, what properties are 'bad'?'.

Part two of the final session starts with a tutor-led discussion of the findings. The aim is for the students to determine which fundamental beliefs about the nature of the environment informed each team and to map each (unnamed) doctrine, at least notionally, across the three major themes. Discussion is informed by reminders about the preceding sections of the course.

As the discussion progresses, the session moves on to more fundamental matters. Since the exercise employs the River Thames, one useful development for the session seems to be to ask each team to reflect on the river: 'According to group A, B, C, D, etc, what is the significance of the river?'. In 1993/4, the student replies suggested that the river was (variously):

1. A total irrelevance, an illusion and a distraction.
2. A shadow of a perfect form.
3. A resource and a challenge (which must be variously developed, marketed or managed).
4. A public utility of varying but substantial value.
5. A mode of transport.

6. A lesson on the problem of the harmonious, sustainable development of society and nature.
7. A lesson on the fall from grace and the corruption of nature by human pride.
8. A physical system controlled by the functional laws of science.
9. An entity, a living being striving for perfection and wholeness.

Further discussion reveals that the way the environment is valued depends on the way it is originally conceived. The environment, conceived as being 'like a machine', is valued in different terms to that same environment conceived as being 'like an organism' or 'like a spiritual experience'. Equally, where the environment is conceived as being 'like a book written by God' it is valued differently to that same environment conceived as being a God, a gift from God, an extension of the internal spirit which is God, or an inanimate material object (Mills 1982). Inevitably, materialistic systems conceive of the environment as a reservoir of things which may be used for the benefit of humankind. Since the different objects of nature offer different utility then they may also be valued, costed in monetary terms (Mitchell and Carson 1989). However, it is not possible to ascribe a monetary value to eternal salvation or inner enlightenment, and it would be a crass sacrilege to place a price on something which is God or a part of God's special personal revelation.

Any assessment of the worth of the environment is affected by the observer's view of the place and purpose of the human species (cf Tucker and Grim 1993 and Box 14.4). Is humankind the crowned ruler of nature, a competitor of nature, an important part of the natural order, a very minor part of nature, a disease of the natural system, the servant, steward or shop steward of nature (cf Lovelock 1991). Is the ultimate purpose of humanity to worship God, to seek individual enlightenment, to promote the sustainable development of society and nature, to have dominion over creation, to act as world guardians, or to reproduce the biosphere on other worlds (cf Miller 1989; Doctor 1967; Frolov 1983; Leiss 1974; Sagan 1989)?

Finally, the concluding seminar moves towards some of the ethical and metaphysical issues raised by the exercise. These include the culturally specific meanings of concepts of nature and value; contrasting definitions of the relationship between humans and the physical world; the appropriateness of some more or less familiar 'world views' as a basis for environmental decision making; the role of public and private world view differences of individuals and groups in conflict generation; and the extent to which it is possible to see things from the perspective of another's culture.

CONTEXT AND STUDENT EVALUATIONS

This exercise could be adapted for application in any social or environmental science curriculum. However, it has been developed to provide a final polish for

Box 14.4   Comparing sarvodaya (Gandhian), ecosocialist (Marxian), Christian fundamentalist and Aquarian ('New Age') environmental viewpoints

The world is:

Sarvodaya: part of the universal spirit or atman.
Ecosocialist: part of the material means of production.
Christian Fundamentalist: the creature of God.
Aquarian: a self-organising, organism, Gaia.
Conventional Western: a resource to be developed.

Truth is:

Sarvodaya: Truth (Satya) is God.
Ecosocialist: concrete but partial – an aspect of a particular dialectical thesis/antithesis.
Christian Fundamentalist: the written word of God recorded in the Bible.
Aquarian: personal subjective enlightenment.
Conventional Western: objective, incontrovertible and immutable fact.

Humans are:

Sarvodaya: part of nature and part of the universal spirit.
Ecosocialist: part of a materialist dialectical relationship between society and its environment.
Christian Fundamentalist: created in God's own image as stewards/managers of the Earth.
Aquarian: aspects of nature and the world spirit (Noosphere).
Conventional Western: apart from nature, they are masters and owners of the world, individual units of economic production and decision making rankable in terms of productivity and accumulated wealth.

Nature is:

Sarvodaya: a divine, all-encompassing creation, of great spiritual purity, which humans should try to experience directly, and live within harmoniously, by creating a political economy that limits 'wants' to the surplus of the immediate environment.
Ecosocialist: something which should be scientifically, rationally, and harmoniously developed for the benefit of society.
Christian Fundamentalist: God's perfect and cared-for creation, given in trust to humanity for its use and instruction.
Aquarian: an harmonious, all-embracing spiritual and material wholeness that includes humanity and is all too often corrupted by human insensitivity and ignorance.

Conventional Western: an economic resource to be exploited in support of economic growth and technological progress and to be preserved in museums/reserves for the purposes of scientific study, general education, and in order to protect such as may prove valuable for future development and exploitation.

(*Sources*: Sarvodaya: Doctor 1967; Ecosocialism: Pepper 1993; Christian Fundamentalist and Aquarian: Miller 1989; Conventional: Metzner 1993).

60–90 students in their final term reading geography within the modular course at Oxford Brookes University. (Those specifically interested in the geography-specific context and application of this exercise or who wish to see sample briefing sheets and other illustrative materials should refer to Haigh et al 1995).

However, in brief, Brookes' geography students read geography as one of two major fields of study. The subject is commonly combined with subjects ranging from humanities such as English and music, social sciences such as anthropology and sociology, sciences including biology and environmental sciences, through technological subjects such as cartography and computer science to vocational specialisations in business and planning.

The programme consciously presents a non-traditional view of geography. The subject is treated as a single, integrated discipline with a focus on the interactions between society and environment. This sidelines several common alternative approaches which conceive of geography as the science of spatial distributions, the study of places and region, or the study of landscape and landscape-forming processes. The programme tries to downplay major divisions in the discipline which divide the several fragments of 'physical geography' from those in 'human geography' and from each other.

Brooke's geography programme is also taught in a non-traditional fashion. The geography team prefer teaching methods which promote active learning and minimise the number of occasions when students may relax as passive recipients of knowledge (cf Ramsden 1992). From the very start of the course, students are habituated to the notion that they will have to collect and collate their knowledge outside the classroom, that they will have to examine and explain their personal understandings in front of the class as a whole, that they will be expected to engage in role-play exercises, and that they should be alert both to designer bias in the texts they are assigned and to a wider purpose in the tasks they perform (cf Jenkins et al 1993; Haigh and Gold 1993). In sum, the exercise described here is by no means an isolated event but forms part of the broad run of teaching, and this may be important for its student acceptability.

In the first year of this exercise, there was an underestimation of the unusual stresses faced by students in their final term of the modular course credit accumulation system who found their overall grades close to one of the boundaries for degree classification. Understandably, such students are in no mood for adventure. However subsequently, as the 'Landscape Assay' became part of student folklore and the students built experience from what had won good grades in the past, this type of resistance has faded.

Today, I remain happy with the exercise. The drama of the final session seems to provide a memorable conclusion to the course. The unpredictable, often theatrical attributes of the session make for a very enjoyable experience. Last, and not least, the two-stage session presents a great challenge for the tutor and, when its course is run, when the session has advanced from its student presentation raw materials through to its final synthesis, the reward is a considerable sense of achievement.

CONCLUSIONS

The 'Landscape Assay' is an academic exercise which aims to immerse students in the implications of cultural diversity and philosophical pluralism for environmental prescription. Since direct experience ranks among the best routes to deep understanding, this exercise involves students in active learning (Ramsden 1992). Students engage in tasks which require them first to see the world through the eyes of the followers of an alien philosophical system, and later to use their own experience of representing an alien world view to see beneath the surface of pronouncements from those who build on different foundations.

Final-year undergraduates have spent three or more years becoming acculturated to the rules, values and procedures of an academic institution. For the most part, they have absorbed the dominant ethos of their university. Those few who have become involved in environmental action will have done so from a very limited perspective. It is unusual for them to have considered the ethics of their actions or to have to examined their actions from the perspectives of another value system. After graduation students often find themselves within organisations which have a more focused, if usually less rationalised, approach to the environment. In this new culture, the functional preconceptions and traditions on which actions are defined are unlikely to be discussed. It is important for the newcomer to be able to recognise what they are. Frequently, students find themselves in organisations involved in conflict or competition with other groups. These actions may be handled more effectively if the newcomer is also able to see through to the beliefs that underlie another's viewpoint and recognise the nuances which indicate the alien value system which they construct.

The Landscape Assay programme seeks to enhance a student's sensitivity to the implications of cultural pluralism. It also attempts to give students the skills needed to recognise, appreciate and comprehend the character of some unfamiliar value systems.

REFERENCES

Capra, F. (1982) *The Turning Point: Science, Society and the Rising Culture*, Collins, London.

Doctor, A. H. (1967) *Sarvodaya*, Asia, New Delhi.

Frolov, I. (1983) 'The Marxist–Leninist Conception of the Ecological Problem', in A. D. Ursul (ed) *Philosophy and the Ecological Problems of Civilization*, Progress, Moscow, pp34–57.

Gold, J. R. (1994) 'Society, Nature, and the Origins of Modern Environmental Thought', unpublished M2662 course workbook, Oxford Brookes University.

Gold, J. R., Haigh, M. J. and Revill, G. (1995) 'Interpreting Disaster: Learning about Environmental Philosophy from *The Grapes of Wrath*', unpublished, Oxford Brookes University.

Haigh, M. J. (1993) 'World Hypotheses: the Origins of Environmental Value Systems', in Y. Guerrier (ed) *Values and Environment: Proceedings of the Conference*, pp30–35, Guildford, University of Surrey.

Haigh, M. J. (1994) 'World Hypotheses Workbook', unpublished M2662 course workbook, Oxford Brookes University.

Haigh, M. J. and Gold, J. R. (1993) 'The Problems with Fieldwork: a Group-based Approach Towards Integrating Fieldwork into the Undergraduate Geography Curriculum', *Journal of Geography in Higher Education*, **17**, 21–32.

Haigh, M. J., Revill, G. and Gold, J. R. (1995) 'The Landscape Assay: Exploring Pluralism in Environmental Interpretation', *Journal of Geography in Higher Education*, **19**, 41–55.

Jenkins, A., Gold, J. R. and Haigh, M. J. (1993) 'Values in Place: the Use of a Journalism Simulation to Explore Environmental Interpretation', in F. Percival, S. Lodge and D. Saunders (eds) *The Simulation and Gaming Yearbook 1993 (Developing Transferable Skills in Education and Training)*, pp211–17, SAGSET/Kogan Page, London.

Kerr, R. S. (1960) *Land, Wood and Water*, Fleet, New York.

Kuhn, T. S. (1970) *The Structure of Scientific Revolutions* (2nd edn), University of Chicago Press, Chicago.

Leiss, W. (1974) *The Domination of Nature*, Beacon Press, Boston.

Lovelock, J. E. (1991) *Gaia: The Practical Science of Planetary Medicine*, Gaia Books, Stroud.

Metzner, R. (1993) 'Emerging Ecological Worldview' (revision of 'The Age of Ecology', *Resurgence*, **149**, 1991) *Bucknell University Review*, **37**, 2, 163–72.

Miller, E. (1989) *A Crash Course on the New Age Movement*, Monarch, Eastbourne.

Mills, W. J. (1982) 'Metaphorical Vision: Changes in Western Attitudes to the Environment', *Association of American Geographers, Annals*, **72**, 2, 237–53.

Mitchell, R. C. and Carson, R. T. (1989) *Using Surveys to Value Public Goods: the Contingent Valuation Method*, Resources for the Future, Washington, DC.

Pepper, D. M. (1993) *Ecosocialism*, Routledge, London.

Pepper, S. C. (1942) *World Hypotheses*, University of California, Berkeley.

Pojman, L. (ed) (1994) *Environmental Ethics: Readings in Theory and Application*, Jones and Bartlett, Boston.

Ramsden, P. (1992) *Learning to Teach in Higher Education*, Routledge, London.

Rorty, R. (1980) *Philosophy and the Mirror of Nature*, Blackwell, Oxford.

Sagan, D. (1989) *Biospheres*, Routledge-Arkana, London.

Shiva, V. (1993) 'Chipko's Earth Charter', in G. S. Rajwar (ed) *Garhwal Himalaya: Ecology and Environment*, pp241–50, Ashish, New Delhi.

Tagore, R. (1965) 'The Wrath of the Gods', in A. Chakravarty (ed) *Rabindranath Tagore: – the Housewarming and other Selected Writings*, pp307–11, Signet New American Library, New York.

Tucker, M. E. and Grim, J. (eds) (1993) *Worldviews and Ecology*, Bucknell University Press, Lewisburg, Pa.

Van Inwagen, P. (1993) *Metaphysics*, Oxford University Press, Oxford.

Author Index

Subject Index